AF411332

Value of Mineralogical Monitoring for the Mining and Minerals Industry

Value of Mineralogical Monitoring for the Mining and Minerals Industry

Editors

Herbert Pöllmann
Uwe König

MDPI • Basel • Beijing • Wuhan • Barcelona • Belgrade • Manchester • Tokyo • Cluj • Tianjin

Editors
Herbert Pöllmann
Martin-Luther-University
Halle-Wittenberg
Germany

Uwe König
Malvern Panalytical B.V.
Almelo
The Netherlands

Editorial Office
MDPI
St. Alban-Anlage 66
4052 Basel, Switzerland

This is a reprint of articles from the Special Issue published online in the open access journal *Minerals* (ISSN 2075-163X) (available at: https://www.mdpi.com/journal/minerals/special_issues/WM4I).

For citation purposes, cite each article independently as indicated on the article page online and as indicated below:

LastName, A.A.; LastName, B.B.; LastName, C.C. Article Title. *Journal Name* **Year**, *Volume Number*, Page Range.

ISBN 978-3-0365-4893-7 (Hbk)
ISBN 978-3-0365-4894-4 (PDF)

Contents

About the Editors

Herbert Pöllmann

Herbert Pöllmann (born 1956 in Waldsassen, † 2022 in Halle (Saale)) studied mineralogy at the Friedrich-Alexander University of Erlangen-Nuremberg where he received his Ph.D. in 1984. In 1994, he was appointed Professor of Mineralogy and Geochemistry at the Institute for Geosciences and Geography at the Martin Luther University in Halle (Saale), Germany. Herbert Pöllmann headed the research group for mineralogy and geochemistry in Halle, supervised numerous Ph.D. theses, and was very extensively involved in research. This research included investigating cementitious mineral formations, the immobilization of pollutants in storage minerals, and saving carbon dioxide by replacing limestone during cement production (through pozzolan, in particular). He also studied mineralogical–geological topics, including mineral raw materials and secondary minerals in basalts. Herbert Pöllmann discovered and worked on five new minerals—kuzelite (1997), serrabrancaite (2000), fluoronatromicrolite (2011), lagalyite (2017) and freitalit (2020)—as the main or co-author. In May 2022, Herbert Pöllmann died completely unexpectedly, shortly before his retirement.

Uwe König

Uwe König studied geology and applied mineralogy at Martin-Luther-University, Halle-Wittenberg, Germany, and the UFPA (Universidade Federal do Para), Belem, Brazil. In 2006, he obtained his Ph.D. in the field of applied mineralogy. Uwe König has worked for Malvern Panalytical B.V. in the Netherlands since 2005, and is responsible for the development of new analytical solutions for the mining, minerals and metals industry. During his career, he has worked as an application specialist, channel partner manager, product manager, and segment manager. His research interests include the use of layered double hydroxides (LDHs) for the immobilization and storage of compounds in industrial residues, the use of mineralogical and geometallurgical monitoring during mining and ore processing, and the combination of different analytical sensors to predict the behavior of ores during beneficiation and processing.

Preface to "Value of Mineralogical Monitoring for the Mining and Minerals Industry"

This Special Issue of *Minerals* provides an overview on the value of mineralogical monitoring during the mining and processing of ores and minerals. Selected case studies from "green metals", e.g., lithium ores, nickel laterites, bauxite, copper ores, and heavy mineral sands, highlight how frequent and accurate mineralogical monitoring has become a standard tool to monitor geometallurgical properties, increase recovery rates, and boost energy efficiency. Most contributions use X-ray diffraction (XRD) as an industrial sensor to identify and quantify mineralogical composition and monitor process parameters. Recently, statistical methods such as cluster analysis and partial least squares regression (PLSR) in combination with XRD raw data have become increasingly popular for handling large amounts of data and correlating process-relevant parameters directly with XRD measurements. This next generation of information, analytical sensors and data enables the use of artificial intelligence (AI) and machine learning (ML) in the mining industry to enhance performance and efficiency.

Herbert Pöllmann and Uwe König
Editors

Editorial

Value of Mineralogical Monitoring for the Mining and Minerals Industry

Uwe König [1],* and Herbert Pöllmann [2],†

[1] Malvern Panalytical B.V., 7602 EA Almelo, The Netherlands
[2] Department of Mineralogy, Institute of Geosciences and Geography, Martin-Luther Halle-Wittenberg University, 06108 Halle, Germany
* Correspondence: uwe.koenig@malvernpanalytical.com
† Herbert Pöllmann passed away during the preparation of this Editorial.

Citation: König, U.; Pöllmann, H. Value of Mineralogical Monitoring for the Mining and Minerals Industry. *Minerals* **2022**, *12*, 902. https://doi.org/10.3390/min12070902

Received: 6 July 2022
Accepted: 8 July 2022
Published: 19 July 2022

Publisher's Note: MDPI stays neutral with regard to jurisdictional claims in published maps and institutional affiliations.

The shift towards lower grade ore deposits, sustainable energy, CO_2 reduction, volatile market conditions and digitalization has pushed the mining and minerals industry towards predictive, sustainable and agile analytical solutions to improve safety and increase operational efficiency. Therefore, fast and frequent mineralogical monitoring, geometallurgical modeling, and the prediction of process-related parameters provides value for mining operations.

Traditionally, quality control in mining industries has relied on time-consuming wet chemistry and on the analysis of elemental composition. However, the mineralogy ruling the physical properties of an ore is often monitored infrequently (if at all). The use of high-speed detectors has turned X-ray diffraction (XRD) into an important tool for fast and accurate process control, even for ores with a complex mineralogy. Recent statistical methods such as cluster analysis or partial least squares regression (PLSR) in combination with XRD raw data have become increasingly popular for handling large amounts of data and correlating process-relevant parameters directly with XRD measurements [1]. The next generation of information processing, new analytical sensors, and big data enables the use of artificial intelligence (AI) and machine learning (ML) in the mining industry to further enhance performance and efficiency.

Mineralogical monitoring is already the standard method to control and monitor processes in other industries, such as cement manufacturing [2] and aluminum smelting [3]. This Special Issue of *Minerals* demonstrates the value of its applications for the mining and minerals industry. The focus is on so-called "green metals" such as nickel, lithium, copper, aluminum and titanium. These metals will be required for energy transition in the coming years.

Nickel laterite production is on the rise and is surpassing conventional sulfide deposits. The efficiency of mining and processing nickel laterites is defined by their mineralogical composition. Mineralogy plays a key role in the production of nickel metal from nickel laterites. The value of mineralogical monitoring for grade definition, ore sorting, and processing is explained by König [4].

As lithium cannot be analyzed by X-ray fluorescence (XRF), this element is monitored by time-consuming wet chemical methods. The use of XRD for the quantitative analysis of lithium minerals and the recalculation of lithium content using statistical methods is discussed in the paper by Pöllmann and König [5]. In addition to addressing hard rock lithium ore analysis, more complex considerations on how to analyze lithium salt brines are included.

Quantitative XRD as a tool to monitor optimal blending and the detection of penalty minerals—which affect the flotation and concentration quality of copper ores—is described by Pernechele et al. [6]. The use of mineralogical monitoring for real-time decisions is discussed. The paper by Can et al. [7] on copper sulfides demonstrates the influence of pyrite mineralogy on the flotation process, and the possibility of developing alternative conditions to improve the performance of the process.

Available alumina and reactive silica are the main parameters controlling the beneficiation of bauxite, which is traditionally measured by laborious, expensive, and time-consuming wet chemical digestion. Alternative methods based on XRD analysis are evaluated by Melo et al. [8]. The potential of these methods industrially applied for rapid and automated quality control of bauxites is demonstrated. The use of mineralogical analysis of alumina-rich clays—covering the largest and most important bauxitic deposits of northern Brazil— as possible raw materials for the local cement and ceramic industry are discussed by Negrao et al. [9].

Heavy mineral sands are the source of various commodities, such as white titanium dioxide pigment and titanium metal. König and Verryn [10] provide information about the use of XRD to determine the composition of raw ores, heavy mineral concentrates, and titania slag. The paper highlights the importance of the fast and direct analysis of the phase composition due to the fact that the efficiency of the different process steps depends on the exact composition of the various titanium and iron phases and the different oxidation stages.

Otoijamun et al. study barite from selected locations in Nigeria [11] and aim to determine its suitability for various industrial applications. The paper shows the added value of XRD in developing beneficiation procedures, processes, and technologies for barite purification.

A systematic review concerning developing solutions based on machine learning to utilize mineralogical data in mining and mineral studies is given in the paper by Jooshaki et al. [12]. They highlight the importance of high-quality and extensive mineralogical information with respect to the increasing global demand for raw materials and evaluate the complexities of the geological structure of ore deposits and decreasing ore grades.

Author Contributions: Writing—original draft preparation, U.K. and H.P.; writing—review and editing, U.K.

Funding: This research received no external funding.

References

1. Degen, T.; Sadki, M.; Bron, E.; König, U.; Nenert, G. The HighScore suite. *Powder Diffr.* **2014**, *29* (Suppl. 2), S78–S83. [CrossRef]
2. König, U. *XPERT: Rietveld and XRD*; International Cement Review: Dorking, UK, 2008; pp. 118–122.
3. König, U.; Norberg, N. Alternative Methods for Process Control in Aluminium Industries—XRD in Combination with PLSR. In Proceedings of the ICSOBA, Quebec City, QC, Canada, 3–6 October 2016; Available online: https://icsoba.org/proceedings/34th-conference-and-exhibition-icsoba-2016/?doc=58 (accessed on 5 July 2022).
4. König, U. Nickel laterites—Mineralogical Monitoring for Grade Definition and Process Optimization. *Minerals* **2021**, *11*, 1178. [CrossRef]
5. Pöllmann, H.; König, U. Monitoring of Lithium Contents in Lithium Ores and Concentrate-Assessment Using X-ray Diffraction (XRD). *Minerals* **2021**, *11*, 1058. [CrossRef]
6. Pernechele, M.; López, A.; Davoise, D.; Maestre, M.; König, U.; Norberg, N. Value of Rapid Mineralogical Monitoring of Copper Ores. *Minerals* **2021**, *11*, 1142. [CrossRef]
7. Can, I.B.; Özçelik, S.; Ekmekçi, Z. Effects of Pyrite Texture on Flotation Performance of Copper Sulfide Ores. *Minerals* **2021**, *11*, 1218. [CrossRef]
8. Melo, C.C.A.; Angélica, R.S.; Paz, S.P.A. A Method for Quality Control of Bauxites: Case Study of Brazilian Bauxites Using PLSR on Transmission XRD Data. *Minerals* **2021**, *11*, 1054. [CrossRef]
9. Negrão, L.B.A.; Pöllmann, H.; Alves, T.K.C. Mineralogical Appraisal of Bauxite Overburdens from Brazil. *Minerals* **2021**, *11*, 677. [CrossRef]

10. König, U.; Verryn, S.M.C. Heavy Mineral Sands Mining and Downstream Processing: Value of Mineralogical Monitoring Using XRD. *Minerals* **2021**, *11*, 1253. [CrossRef]
11. Otoijamun, I.; Kigozi, M.; Adetunji, A.R.; Onwualu, P.A. Characterization and Suitability of Nigerian Barites for Different Industrial Applications. *Minerals* **2021**, *11*, 360. [CrossRef]
12. Jooshaki, M.; Nad, A.; Michaux, S. A Systematic Review on the Application of Machine Learning in Exploiting Mineralogical Data in Mining and Mineral Industry. *Minerals* **2021**, *11*, 816. [CrossRef]

Article

Nickel Laterites—Mineralogical Monitoring for Grade Definition and Process Optimization

Uwe König

Malvern Panalytical B.V., Lelyweg 1, 7602 EA Almelo, The Netherlands; uwe.koenig@malvernpanalytical.com

Abstract: Nickel laterite ore is used to produce nickel metal, predominantly to manufacture stainless steel as well as nickel sulfate, a key ingredient in the batteries that drive electric vehicles. Nickel laterite production is on the rise and surpassing conventional sulfide deposits. The efficiency of mining and processing nickel laterites is defined by their mineralogical composition. Typical profiles of nickel laterites are divided into a saprolite and a laterite horizon. Nickel is mainly concentrated and hosted in a variety of secondary oxides, hydrous Mg silicates and clay minerals like smectite or lizardite in the saprolite horizon, whereas the laterite horizon can host cobalt that could be extracted as a side product. For this case study, 40 samples from both saprolite and laterite horizons were investigated using X-ray diffraction (XRD) in combination with statistical methods such as cluster analysis. Besides the identification of the different mineral phases, the quantitative composition of the samples was also determined with the Rietveld method. Data clustering of the samples was tested and allows a fast and easy separation of the different lithologies and ore grades. Mineralogy also plays a key role during further processing of nickel laterites to nickel metal. XRD was used to monitor the mineralogy of calcine, matte and slag. The value of mineralogical monitoring for grade definition, ore sorting, and processing is explained in the paper.

Keywords: nickel laterite; ore sorting; XRD; Rietveld; cluster analysis

Citation: König, U. Nickel Laterites—Mineralogical Monitoring for Grade Definition and Process Optimization. *Minerals* **2021**, *11*, 1178. https://doi.org/10.3390/min11111178

Academic Editor: Fang Xia

Received: 21 September 2021
Accepted: 21 October 2021
Published: 24 October 2021

Publisher's Note: MDPI stays neutral with regard to jurisdictional claims in published maps and institutional affiliations.

1. Introduction

Battery manufacturing together with the demand for stainless steel is the biggest driver for the global nickel mining industry. About 60% to 70% of the current worldwide nickel resources are derived from laterites whereas the rest is extracted from nickel sulfide ores [1,2]. However, nickel laterites account currently only for about 40% of the global nickel production. Since ore grades and resources of sulfide nickel deposits generally decrease, mining companies are forced to focus more on the extraction of nickel from laterites in the future, see Figure 1.

Primary nickel production is generally divided into two main product categories. Nickel class I (nickel content > 99%) describes a group of nickel products comprising electrolytic nickel, powders, and briquettes, as well as carbonyl nickel. Nickel class II (nickel content < 99%) comprises nickel pig iron and ferronickel. These nickel products are used especially in stainless steel production. Roughly 48% of the total nickel mining output is related to class I nickel products, with class II nickel products accounting for the remaining 52% [3]. While class II nickel is mainly obtained from laterites, the production of class I nickel is based primarily on sulfide ores but moving to laterites too. Battery production requires class I nickel. Currently, approximately 5% of the world nickel production is used to manufacture batteries. The high demand for electric vehicles together with increasing the energy density of the batteries forces manufacturers more and more to increase the nickel content and to decrease the cobalt content in batteries. In 2025 the demand for nickel for batteries is expected to increase by approximately 15% of the world nickel production [3]. The additional nickel supply will be mined mainly from lateritic deposits.

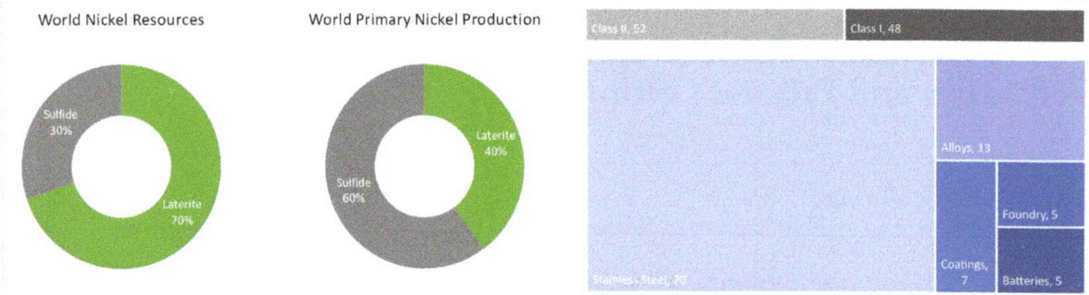

Figure 1. World nickel resources, production, and use (%).

To extract nickel from laterites hydrometallurgical [4] and pyrometallurgical [5–7] processes are used. Whereas high-grade nickel laterites (>2% Ni) are mainly processed pyrometallurgically, lower-grade nickel deposits (<1.3% Ni) are processed mainly hydrometallurgically.

Nickel laterites form under humid tropical conditions during the weathering of serpentinite rocks. The mineralogy and ore grade depend on the lithology and climate during the formation of the deposit. Nickel is hosted in several minerals such as oxides, Mg silicates and clays. Laterite-type resources are found in Indonesia, the Philippines, Brazil, Cuba, Australia and New Caledonia. Lateritic nickel deposits can be classified in mainly three groups [8,9]: (a) oxidic or "limonitic" deposits dominated by minerals such as goethite FeOOH, (b) smectitic or "clay mineral" deposits dominated by nickel-bearing swelling clays such as smectite or nontronite and (c) hydrous Mg-Si-silicate deposits dominated by talc- and serpentine-like minerals, collectively referred to as "garnierites" [10,11] that occur in the saprolite zone of the yellow laterite profile, Figure 2.

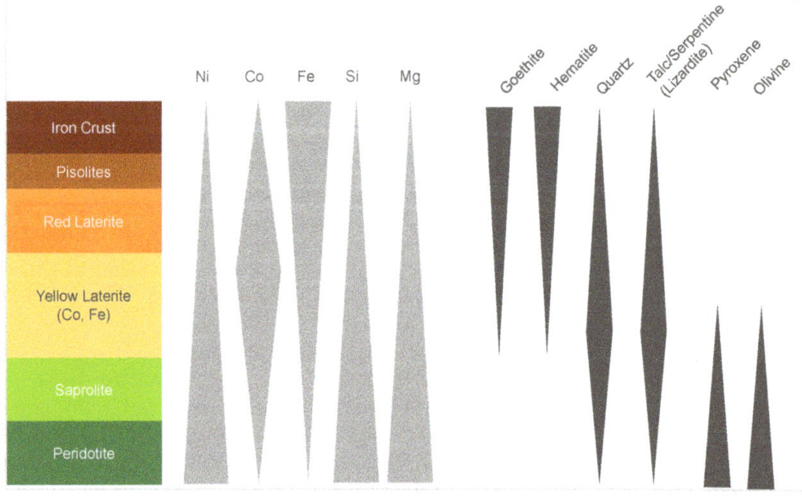

Figure 2. Schematic hydrous Mg-Si-silicate laterite profile modified after [12,13].

Garnierite is a general name defining greenish, poorly crystallized, clay-like Ni ores that generally comprises an intimate mixture of Ni/Mg hydrosilicates like serpentine, lizardite, talc, sepiolite, smectite, and chlorite [10].

Currently, ore grades are mainly defined based on the elemental composition. Mineralogical analysis is only used for research on dedicated samples during mining and processing of nickel laterites or for the exploration of new deposits. This paper provides an overview of the value of mineralogical monitoring of lateritic nickel ores to increase the efficiency and the metal recovery during mining, ore sorting and blending as well as during pyrometallurgical processing.

2. Materials and Methods

2.1. Samples and Sample Preparation

40 nickel laterite samples from New Caledonia plus 6 samples from processed laterites were analyzed for this case study. The 40 samples are from both laterite and saprolite horizons of the nickel laterite profile, Figure 3. To guarantee a reproducible and constant sample preparation for the XRD measurements, the samples were prepared as pressed pellets using automated sample preparation equipment. All powder samples were milled for 30 s and pressed 30 s with 10 tons into steel ring sample holders. No binder was used to prepare the samples.

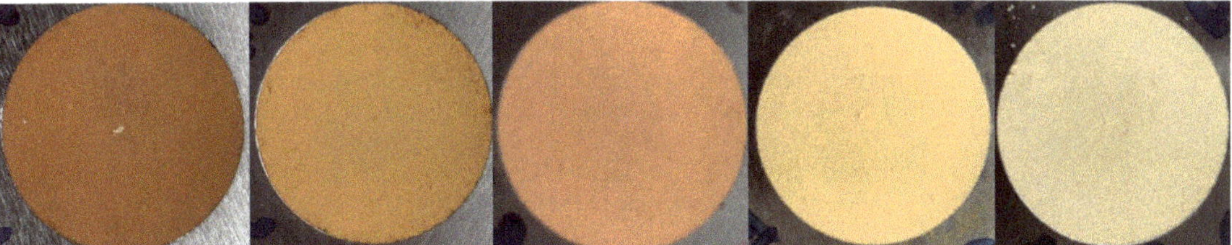

Figure 3. Selected nickel laterite samples prepared for XRD measurements representing five main groups in the nickel laterite profile, left = high goethite, right = high lizardite.

2.2. X-ray Diffraction (XRD)

X-ray powder diffraction (XRD) is a versatile, non-destructive analytical method for the identification and quantitative determination of crystalline phases present in powdered and bulk samples. For the studies presented in this paper, a Malvern Panalytical "Aeris Minerals" benchtop diffractometer (Almelo, The Netherlands) with a cobalt anode, an incident iron filter and a linear detector was used, featuring measurement times of about 5 min per sample. The XRD patterns were collected in the range 5° to 82° 2θ. The setup consists of an X-ray source, a spinning sample stage for optimizing counting statistics, a high-speed detector, and several optics.

Data evaluation was performed with the software package HighScore Plus version 4.9 [14]. The identification of all crystalline mineral phases is achieved by comparing measured diffraction data to a reference database. For this study, the Crystallography Open Database (COD) from 2021 was used [15].

2.3. Rietveld Quantification

The mineral quantification of all samples was determined using the Rietveld method [16–18]. Modern XRD quantification analysis techniques such as Rietveld analysis are attractive alternatives to classical peak intensity or area-based methods since they do not require any standards or monitors. The method offers impressive accuracy and speed of analysis. The knowledge of the exact crystal structure of all minerals present in the nickel laterite samples is mandatory for the Rietveld refinements.

2.4. Cluster Analysis

To handle large amounts of data achieved by rapid data collection using a linear detector, "cluster analysis" is a useful tool to combine different XRD measurements (and thus different ore grades) into similar groups (clusters) [19,20]. The method can be used for stockpiling different grades of the nickel laterite profile with different mineralogical properties and thus varying process behavior. Cluster analysis can be also used to automatically apply dedicated Rietveld runs on different groups of samples with different mineralogical contents, to improve the accuracy of the quantitative results.

Cluster analysis greatly simplifies the analysis of large amounts of data. It automatically sorts all (closely related) scans of an experiment into separate groups and marks the most representative scan of each group as well as the most outlying scans within each group. Cluster analysis is basically a three-step process, but it contains optional visualization and verifications steps as well:

1. Comparison of all scans in a document with each other, resulting in a correlation matrix representing the dissimilarities of all data points of any given pair of scans.
2. Agglomerative hierarchical cluster analysis puts the scans in different classes defined by their similarity. The output of this step is displayed as a dendrogram, where each scan starts at the left side as an individual cluster. The clusters amalgamate in a stepwise fashion until they are all united in one single group.
3. The best possible grouping (=number of separate clusters) is estimated by the KGS test [19] or by the largest relative step on the dissimilarity scale. This number can be adapted manually too. Additionally, the most representative scan and the two most outlying scans within each cluster is determined and marked.
4. Next to hierarchical clustering you can use three independent tools, namely Principal Components Analysis (PCA), Metric Multi-Dimensional Scaling (MMDS), or t-Stochastic Neighbor Embedding (t-SNE) to define clusters; they are all shown in pseudo-three-dimensional plots.

t-SNE [21], as used in this case study, is a separate and independent method to visualize and to judge the quality of the clustering. Either the correlation matrix of step 1, or the raw data is used as input, the output is again a pseudo-3-dimensional plot.

2.5. Fuzzy Clustering

Cluster analysis is not only a data reduction tool; it can also be used to discover hidden patterns in data as well as expose phase relationships in large numbers of patterns of complex mixtures. In order to be able to deal with phase mixtures without prior knowledge of the possible constituents, fuzzy clustering can be applied to the samples [22,23]. This cluster validation technique allows a member to join more than one cluster. It is sometimes called soft clustering too. For each member the probability (between one and zero) to join every cluster is calculated. The results are shown in a table.

- Probabilities < 0.2 indicate members, which surely do not belong to this cluster.
- Probabilities > 0.7 indicate members, which certainly do belong to a specific cluster.
- Probabilities between 0.2 and 0.7 indicate members, which could belong to more than one cluster. These should be inspected in more detail.

2.6. X-ray Fluorescence (XRF)

Elemental analysis was determined using X-ray fluorescence (XRF) technology. The powder samples were first dried in an oven at 105 °C overnight. The mixture with 12 g dried sample material plus 3 g binder was ground with tungsten carbide swing mill for 30 s. The mixture was then pressed into a 40 mm diameter pellet under 20 tons for 30 s. A Malvern Panalytical "Epsilon 4" bench-top spectrometer (Almelo, The Netherlands) with Rh tube was used with measurement times of about 3 min per sample. Secondary Ni ore standards from New Caledonia were used to setup the calibration.

3. Results and Discussion

Evaluation of the XRD measurements was done in several steps. As a first step data clustering was applied to define mineralogical domains within the laterite horizon. A second step included minerals identification and quantification, and the results were compared to a typical nickel laterite horizon. Finally, the XRD results were validated with the elemental composition determined by XRF.

In addition to the lateritic raw material 6 processed samples were analyzed to identify and quantify the phase composition.

3.1. Cluster Analysis of Nickel Laterite

All 40 samples of the lateritic profile were used for cluster analysis to define groups of similar mineralogical composition. Based on the correlation matrix the dendrogram in Figure 4 defines 4 cluster plus 3 outliers. The different cluster are visualized in a 3-dimensional t-SNE plot, Figure 5.

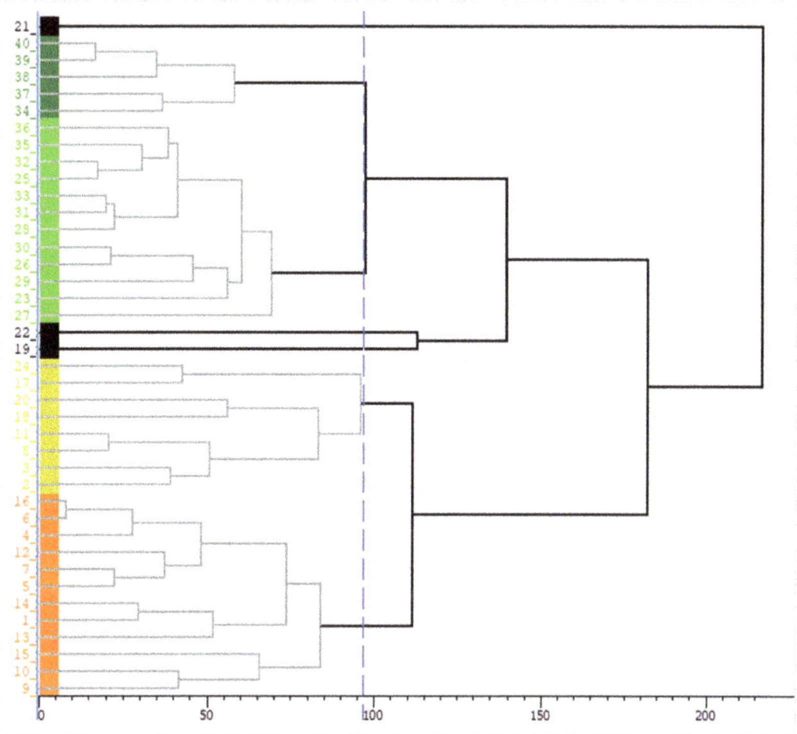

Figure 4. Dendrogram after cluster analysis based on the correlation matrix from 40 nickel laterite samples (cluster 1 = orange, cluster 2 = yellow, cluster 3 = light green, cluster 4 = dark green), X-axis shows the dissimilarity of the tie bars.

Phase identification and quantification, Section 3.2, confirmed that the 4 cluster represent different parts of the laterite horizon. Cluster 1 represents samples from the laterite horizon with high content of goethite whereas cluster 2 contains laterite samples with high quartz content. Clusters 3 and 4 consist of saprolite samples with high lizardite content in cluster 3 and high olivine and pyroxene content in cluster 4.

The outliers are samples with very low goethite content (<10%, samples 19, 21, 22) and with a far higher amount of gibbsite (sample 19) compared to the other samples.

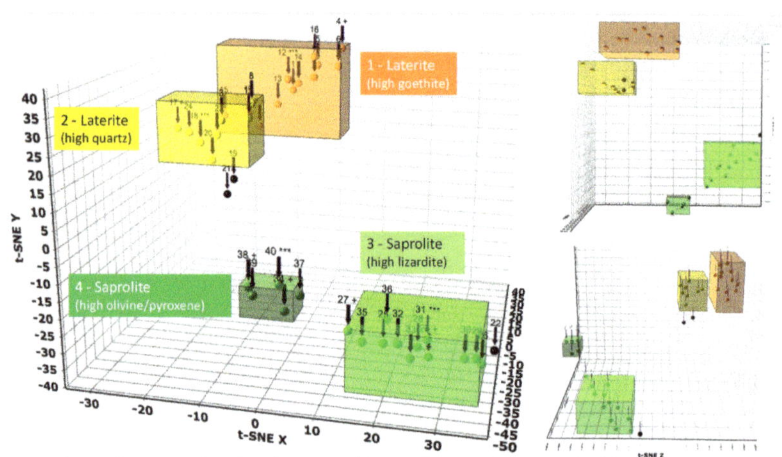

Figure 5. 3D t-SNE score plot after data clustering of 40 samples, visuals from 3 different angles.

To explore the possible transitions between the different mineralogical horizons (in contrast to sharp borders) fuzzy clustering was applied. The fuzzified 3D t-SNE score plot, Figure 6, clearly points towards such transitions. Mixed colors mark the scans that belong to transitions and larger spheres indicate those scans where the membership coefficient exceeds/falls under a certain threshold.

The results in Table 1 show the matrix notation, the so-called membership matrix M, with all calculated probabilities for each measurement. Clusters are organized in columns, probabilities for each sample are shown in a row.

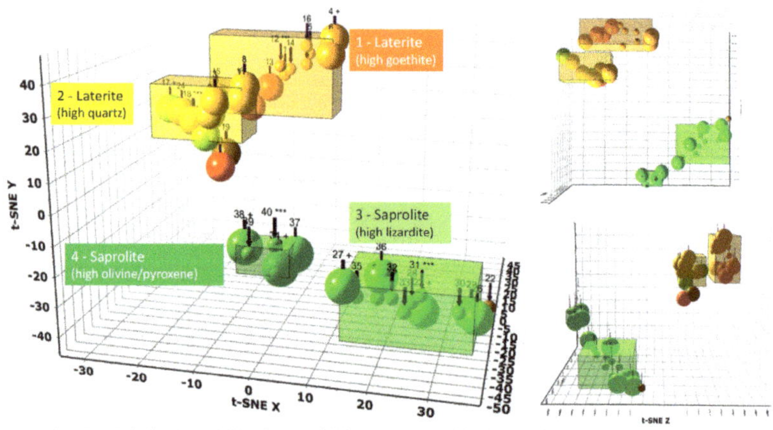

Figure 6. Fuzzified 3D t-SNE score plot, phase mixtures are indicated by mixed colors and larger spheres.

The results prove the capability of fuzzy clustering to detect transitions as mixtures of adjacent clusters. Figure 7 shows the 4 cluster and the membership coefficients. The trendlines indicate high membership and better separation for clusters 1, 3 and 5. Cluster 2 with lower coefficients is made up by mixing mineral associations from cluster one and three and represents transition mineralogy.

Table 1. Results of fuzzy clustering of 40 nickel laterite samples, refined membership coefficients, colors represent the different cluster as identified in Figures 5 and 6, Liz = Lizardite, Oli = Olivine, Ens = Enstatite.

Cluster No.	Sample No.	Laterite	Transition	Saprolite (High Liz)	Saprolite (High Oli/Ens)	Mixture *
1	1	0.8	0.2	0.1	0.1	
2	2	0.4	0.8	0.2	0.0	X
2	3	0.5	0.7	0.1	0.1	X
1	4	0.7	0.4	0.2	0.0	
1	5	0.8	0.3	0.1	0.1	
1	6	0.8	0.4	0.1	0.0	
1	7	0.9	0.4	0.1	0.1	
2	8	0.6	0.7	0.1	0.0	X
1	9	0.7	0.1	0.3	0.1	
1	10	0.8	0.2	0.2	0.1	
2	11	0.5	0.7	0.1	0.1	X
1	12	0.8	0.4	0.2	0.1	
1	13	0.7	0.2	0.1	0.1	
1	14	0.9	0.3	0.1	0.1	
1	15	0.7	0.3	0.3	0.1	
1	16	0.8	0.4	0.1	0.0	
2	17	0.5	0.4	0.5	0.1	X
2	18	0.3	0.8	0.3	0.1	X
2	20	0.2	0.7	0.3	0.0	X
3	23	0.2	0.3	0.7	0.2	
2	24	0.5	0.5	0.4	0.0	X
3	25	0.2	0.1	0.9	0.3	
3	26	0.1	0.3	0.8	0.2	
3	27	0.1	0.1	0.7	0.4	
3	28	0.2	0.2	0.8	0.2	
3	29	0.1	0.4	0.7	0.1	
3	30	0.1	0.3	0.8	0.2	
3	31	0.2	0.2	0.9	0.2	
3	32	0.2	0.1	0.9	0.3	
3	33	0.2	0.1	0.9	0.2	
4	34	0.1	0.0	0.6	0.6	X
3	35	0.2	0.1	0.8	0.3	
3	36	0.2	0.2	0.8	0.4	
4	37	0.1	0.1	0.6	0.7	
4	38	0.1	0.1	0.4	0.7	
4	39	0.1	0.1	0.4	0.8	
4	40	0.1	0.1	0.4	0.8	

* Indicated by Fuzzy Clustering.

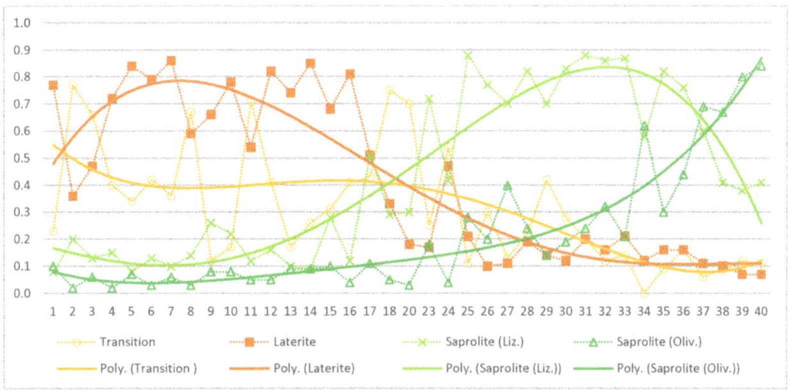

Figure 7. Plot of membership coefficients and trendlines of all clusters against sample number, colors represent the different cluster as identified in Figures 5 and 6 and Table 1.

3.2. Mineral Identification and Quantification

Mineral identification of all 40 nickel laterite samples confirmed the presence of two main groups of ores as determined previously using cluster analysis. One group is dominated by oxidic minerals such as goethite, hematite, gibbsite and quartz whereas the second group is characterized by the presence of residual primary Fe/Mg-silicates such as pyroxene and olivine as well as secondary silicates including lizardite and talc. Lizardite peaks appear broad with an FWHM around 0.3° 2θ which indicates lower crystallinity. Such poorly ordered hydrous equivalents are commonly referred to as "garnierites" [8,12,13,24,25], named after Jules Garnier who first discovered them in 1864 in New Caledonia [26]. It is a generic name for a green nickel ore that has formed by lateritic weathering of ultramafic rocks (serpentinite, dunite, peridotite). Garnierite is mostly a mixture of various Ni- and Ni-bearing magnesium layer silicates and occurs in many nickel laterite deposits in the world [27]. Lizardite is the main nickel-bearing mineral in the analyzed samples. A synonym for Ni-Lizardite is "Népouite" [26]. Expandable clay silicates (e.g., smectite with peaks at ~16 Å and ~19 Å) or (semi)amorphous phases [28,29] were not detected. Structure refinement of the goethite peaks does not show Co-substitution [30–35]. Table 2 summarizes all identified minerals in the investigated nickel laterite samples.

Table 2. Identified minerals in the nickel laterite samples.

Mineral	Formula	References
Goethite	$FeOOH$	[27]
Hematite	Fe_2O_3	[36]
Gibbsite	$Al(OH)_3$	[37]
Quartz	SiO_2	[38]
Lizardite	$(Mg,Ni)_3(Si_2O_5)(OH)_4$	[39]
Talc	$Mg_3[(OH)_2Si_4O_{10}]$	[40]
Enstatite (Pyroxene)	$Mg_{15.44}Ca_{0.56}Si_{16}O_{48}$	[41]
Forsterite (Olivine)	$Mg_{7.17}Fe_{0.8}Ni_{0.02}Mn_{0.01}Si_4O_{16}$	[42]

Figure 8 gives an overview of all XRD measurements from the lateritic profile as a surface plot. Different colors indicate the different intensities of the peak of the minerals. The upper laterite horizon of the profile is dominated by the oxide minerals, while in the lower saprolite horizon secondary and primary silicates are the main minerals.

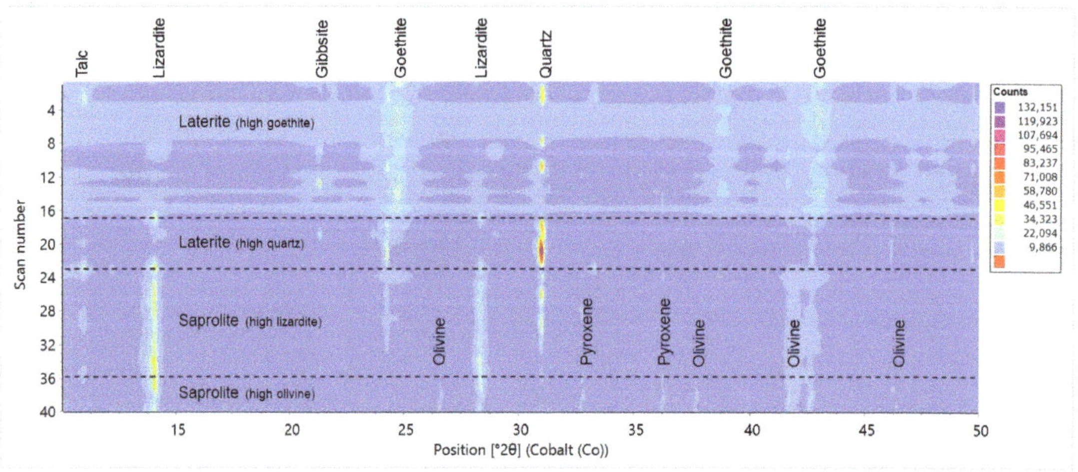

Figure 8. XRD scan surface plot of the region between 10° 2θ and 50° 2θ showing intensities of the main mineral phases.

After mineral identification, the quantitative composition of the samples was determined using the Rietveld method. The Rietveld method is a full-pattern fit method. The measured profile and a profile calculated from crystal structure data are compared. By variation of many parameters, the difference between the two profiles is minimized. Structures and crystallographic data for all phases present in the samples are derived from the COD database, Table 2.

Since refinement depends on finding the best fit between a calculated and experimental pattern, it is important to have a numerical figure of merit quantifying the quality of the fit and to provide insight into how well the model fits the observed data. For this case study, the R_{wp} (weighted profile R-value) was used [16].

Figure 9 shows an example of the resulting full-pattern Rietveld refinement of one lateritic and one saprolitic ore sample. The mineralogical composition and the R_{wp} factors of all samples are summarized in Table 3.

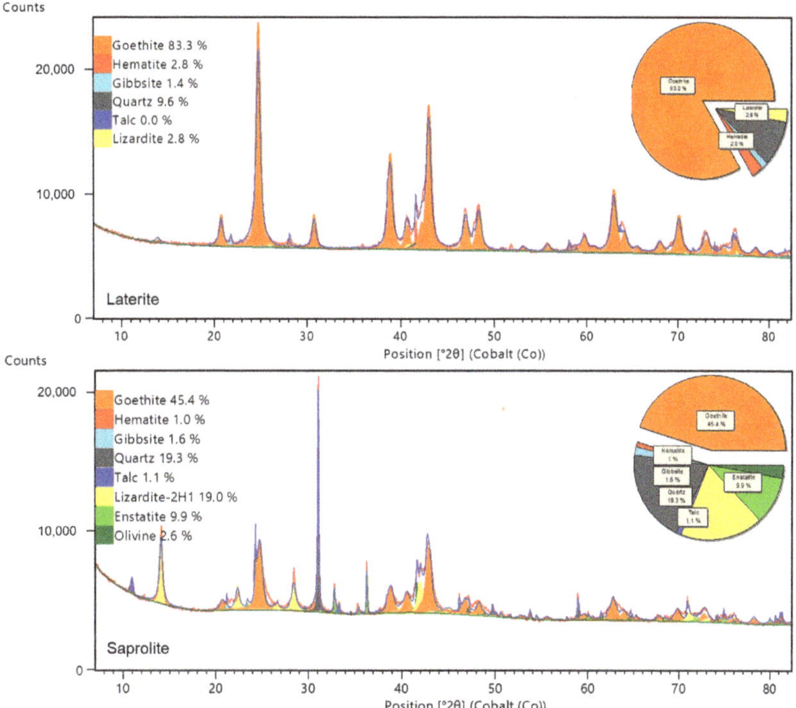

Figure 9. Example of a full-pattern Rietveld quantification of a lateritic (R_{wp} = 3.3) and saprolitic (R_{wp} = 4.6) ore.

After generating a model containing all structures of the expected phases, minerals quantification using the Rietveld method can be applied completely automatic without operator interference. It can be used to collect fast feedback for defining grade blocks in the mine, or to sort and blend ores from different mineralogical domains and laterite horizons for a more homogenous ore. This is important because ore mineralogy directly influences downstream processing as described in Section 3.3. and the nickel recovery rates respectively.

To validate the mineralogical composition obtained with the Rietveld method, the theoretical elemental composition was calculated by breaking down mineral phase concentrations into oxide concentrations, and by comparing them with XRF elemental analysis.

Table 3. Mineral composition determined using the Rietveld method, R_{wp} value of the quantification and main elements analyzed with XRF.

#	Goe * [%]	Hem * [%]	Gib * [%]	Qua * [%]	Tal * [%]	Liz * [%]	Ens * [%]	Oli * [%]	R_{wp}	Ni [%]	Co [%]	Fe$_2$O$_3$ [%]	MgO [%]	SiO$_2$ [%]
1	89.4	2.1	5.6	0.7	0.0	2.2	0.0	0.0	4.6	0.6	0.0	74.7	0.7	1.2
2	49.0	0.7	8.2	33.2	5.3	3.6	0.0	0.0	3.5	1.6	0.1	41.1	4.2	35.8
3	50.4	0.8	4.5	34.3	9.2	0.9	0.0	0.0	4.6	0.7	0.1	46.1	3.6	39.3
4	77.4	0.7	9.1	10.2	0.5	2.1	0.0	0.0	1.9	1.6	0.6	66.3	0.9	2.6
5	83.6	0.3	6.7	7.5	0.2	1.7	0.0	0.0	3.1	1.2	0.2	69.9	0.7	2.1
6	81.6	0.7	6.3	7.7	0.8	3.0	0.0	0.0	2.4	1.9	0.3	66.5	0.9	2.8
7	79.2	1.7	8.3	7.5	0.3	3.0	0.0	0.0	2.7	1.7	0.4	67.2	0.8	2.3
8	63.8	0.3	5.4	27.7	0.2	2.7	0.0	0.0	2.5	1.2	0.2	54.4	0.8	22.2
9	72.5	4.0	7.5	10.4	0.1	5.6	0.0	0.0	6.6	0.7	0.0	59.8	2.9	8.9
10	59.5	3.1	4.9	26.0	0.1	6.3	0.0	0.0	6.9	0.9	0.0	55.5	3.3	19.0
11	47.6	0.4	5.6	44.5	0.2	1.6	0.0	0.0	3.2	0.9	0.1	44.4	0.8	34.2
12	74.2	1.7	11.5	6.4	0.3	5.8	0.0	0.0	2.9	1.4	0.1	64.2	2.6	9.8
13	63.9	3.4	9.5	10.9	2.6	9.7	0.0	0.0	9.6	0.8	0.0	61.7	4.8	12.7
14	83.3	2.8	1.4	9.6	0.0	2.8	0.0	0.0	3.3	0.3	0.0	72.0	0.7	7.1
15	65.7	2.9	8.9	7.2	0.6	14.7	0.0	0.0	4.7	1.1	0.0	56.5	7.1	11.4
16	79.8	0.5	7.6	8.2	0.7	3.2	0.0	0.0	2.3	1.5	0.2	67.1	1.4	6.6
17	85.3	1.3	0.1	4.4	0.6	8.3	0.0	0.0	5.5	0.8	0.1	75.3	4.6	4.5
18	83.8	0.9	0.1	10.4	0.7	4.0	0.0	0.0	8.9	0.9	0.0	71.8	3	11.1
19	9.2	0.5	12.5	45.1	0.1	30.6	1.7	0.1	9.9	2.4	0.1	6.2	16.6	54.9
20	19.6	0.3	0.4	60.4	1.6	15.6	1.9	0.4	6.2	1.2	0.0	12.9	9.4	65
21	5.5	0.2	0.2	86.2	0.1	4.5	2.5	0.7	6.1	0.5	0.0	3.7	5.1	83.4
22	7.8	0.3	0.3	69.9	1.9	19.2	0.5	0.1	8.3	0.9	0.0	4.7	14.2	67.5
23	28.0	1.1	1.8	5.4	25.5	32.4	1.9	3.9	9.4	2.2	0.1	21.2	23.4	37.5
24	45.4	1.0	1.6	19.3	1.1	19.0	9.9	2.6	4.6	1.4	0.0	37.2	15.1	34.9
25	37.2	0.7	1.6	5.8	0.1	40.4	9.3	5.0	7.2	2.6	0.1	27.7	24.7	37.6
26	18.8	0.2	0.5	42.1	0.6	25.0	6.2	6.7	8.5	1.4	0.1	14.7	15.6	62.1
27	25.5	0.1	1.0	21.1	0.3	39.3	6.1	6.6	8.7	2.4	0.0	17.3	20.9	49.9
28	24.3	0.4	0.8	13.2	0.8	26.9	24.9	8.7	8.3	1.9	0.1	17	23.4	48.8
29	21.2	0.5	0.8	29.2	1.9	21.6	14.7	10.2	8.5	1.6	0.0	14.8	21.2	53.6
30	28.4	0.8	0.9	22.2	1.8	34.1	6.9	4.9	7.1	2.6	0.1	23.9	21.0	43.7
31	34.9	1.0	1.5	11.0	0.6	33.4	9.2	8.4	6.9	2.5	0.1	26.4	22.7	38.1
32	35.9	0.7	1.1	8.9	0.5	38.1	7.3	7.4	7.8	2.4	0.1	24.5	23.2	34.1
33	36.5	0.8	1.3	5.5	0.7	39.9	9.3	6.1	7.1	2.3	0.1	26.8	25.3	32.5
34	25.8	0.4	0.2	4.4	0.0	44.7	9.4	15.0	9.7	3.3	0.0	19.5	29.7	39.7
35	26.7	0.3	0.5	6.9	1.8	43.8	11.2	8.7	9.8	3.2	0.0	20.2	28	39.9
36	29.0	0.6	0.6	6.6	1.2	29.7	16.6	15.7	6.9	2.5	0.0	21.2	26.8	37.6
37	21.2	0.0	0.3	2.7	0.7	30.8	12.2	32.1	8.4	2.6	0.0	17.6	31.6	37.2
38	15.0	0.3	0.2	0.7	0.3	18.0	24.8	40.6	7.0	1.5	0.0	15.2	35.7	39.8
39	18.4	0.5	0.0	2.4	0.2	15.7	21.4	41.4	6.3	1.4	0.0	17.1	33.9	43.6
40	18.7	0.2	0.3	1.6	0.3	20.6	18.7	39.6	7.1	1.9	0.0	17.0	33.8	39.9

* Goe = Goethite, Hem = Hematite, Gib = Gibbsite, Qua = Quartz, Tal = Talc, Liz = Lizardite, Ens = Enstatite, Oli = Olivine.

Figure 10 summarized the mineralogical and elemental composition of all samples within the laterite profile. The results for Fe$_2$O$_3$, MgO and SiO$_2$ show a good agreement. The samples analyzed for this case study represent a typical hydrous Mg-Si-silicate laterite profile based on the mineralogical and elemental composition. The goethite dominated iron crust and red laterite are followed by a transition zone with relatively high amounts of quartz. The saprolite is dominated by the increasing amounts of Fe/Mg-silicates such as pyroxene and olivine as well as secondary silicates including lizardite and talc. The high amounts of pyroxene and olives in samples 38–40 indicate proximity to the bedrock of the deposit.

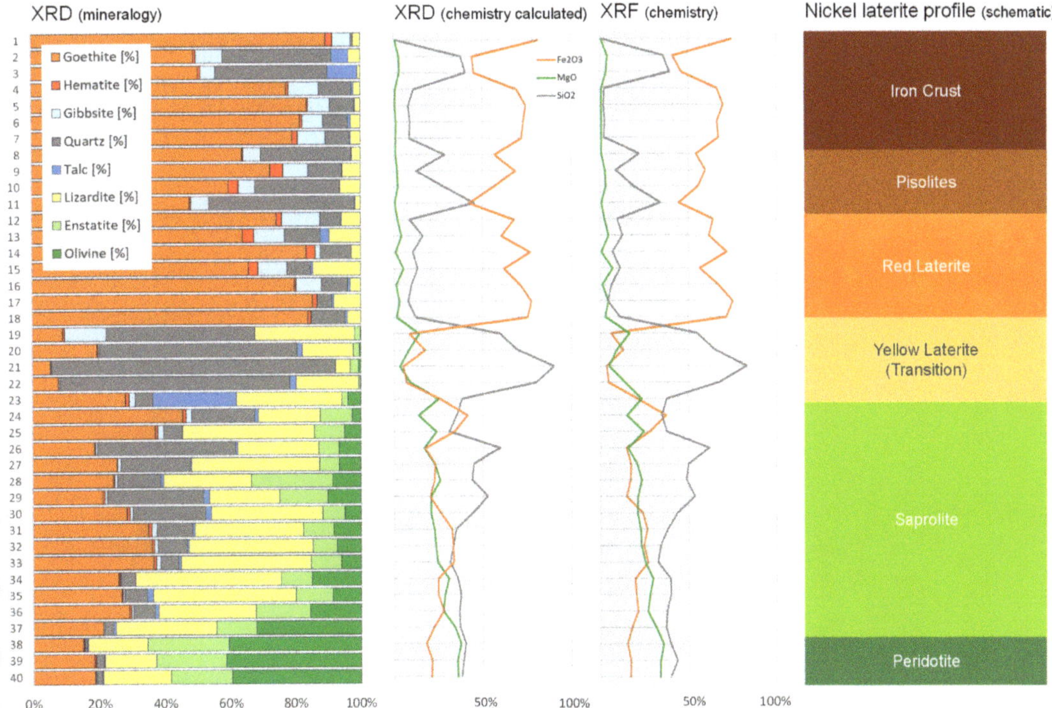

Figure 10. Summary of the mineralogical and elemental composition of 40 samples from a hydrous Mg-Si-silicate laterite profile.

Lizardite is the nickel-bearing mineral in the saprolite and therefore of special economic interest. A comparison of the lizardite and the elemental nickel concentration, Figure 11, shows a good correlation between both values. It proofs that nickel is indeed contained in the mineral lizardite and shows that XRD is beneficial to track not only different ore grades and mineralogical domains but can also give fast feedback about the main nickel mineral and indirectly about the total nickel concentration.

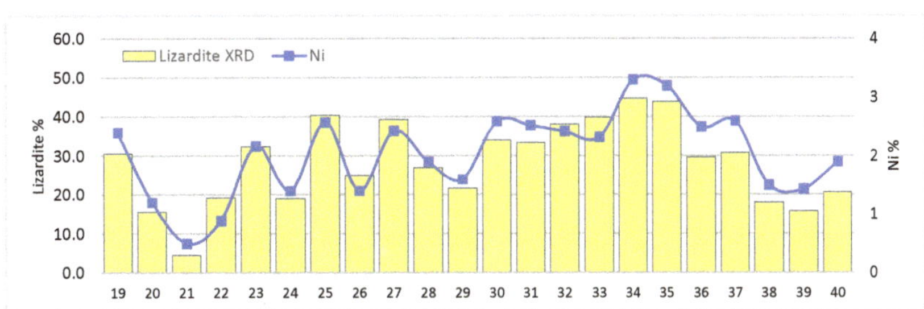

Figure 11. Comparison of the amount of lizardite (XRD) and the nickel content (XRF) in the saprolitic samples of the profile.

3.3. Mineralogical Monitoring during Pyrometallurgical Processing of Nickel Laterites

To ensure optimal efficiency for both mining and ore processing, knowledge about the mineralogical composition of the ore feed, but also of the processed materials and products is mandatory. Varying mineralogical composition of the ore blends can affect the lifetime of refractories, melting temperature, reducibility and recovery rates during pyrometallurgical processing.

Nickel laterites are suited for pyrometallurgical processing involving drying, calcination/reduction and electric furnace smelting to produce ferronickel or nickel sulfide matte. Figure 12 shows a schematic overview of nickel extraction from laterites by smelting them in an electrical furnace.

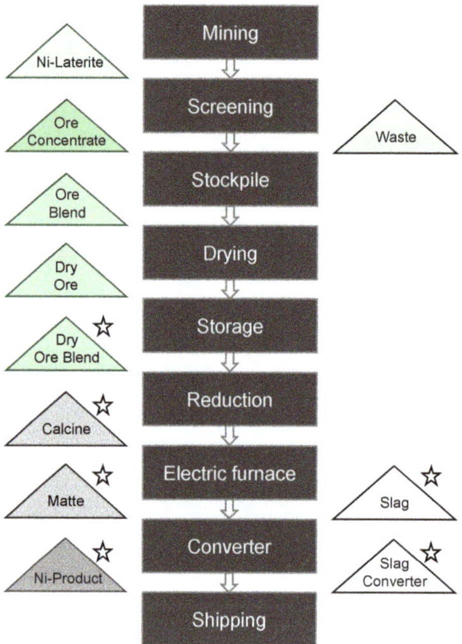

Figure 12. Schematic overview of the pyrometallurgical extraction of nickel from laterites, stars = samples analyzed.

Nickel matte production starts after the ore screening with the drying to reduce the water content from 30–40% to 20%. The dried nickel ore is further treated in a reduction kiln to remove the final water content. The product of this process is called calcine. Calcine is melted and reduced in an electric arc furnace producing a sulfidic matte and an oxidic slag. The slag is skimmed from the furnace continuously and is disposed of. The matte is tapped periodically as required by the converters. Molten furnace matte is transferred to the converters through ladles. Air/oxygen is blown in to oxidize the remaining iron. Silica flux is added to melt the oxidized iron that is then incorporated into the slag. During converting, the lower grade electric furnace matte is converted to Bessemer matte [43]. The final converter product is granulated, dried, screened and packed for shipment.

It is important to control not only the mineralogical composition of the run-of-mine ore but also the blended feed material, calcine, matte and slag. Samples from six different materials were analyzed for the mineralogical composition. All identified phases are summarized in Table 4. Figure 13 gives an overview of the phase composition investigated with the Rietveld method of the different process streams from the ore blend to the Bessemer matte and slag.

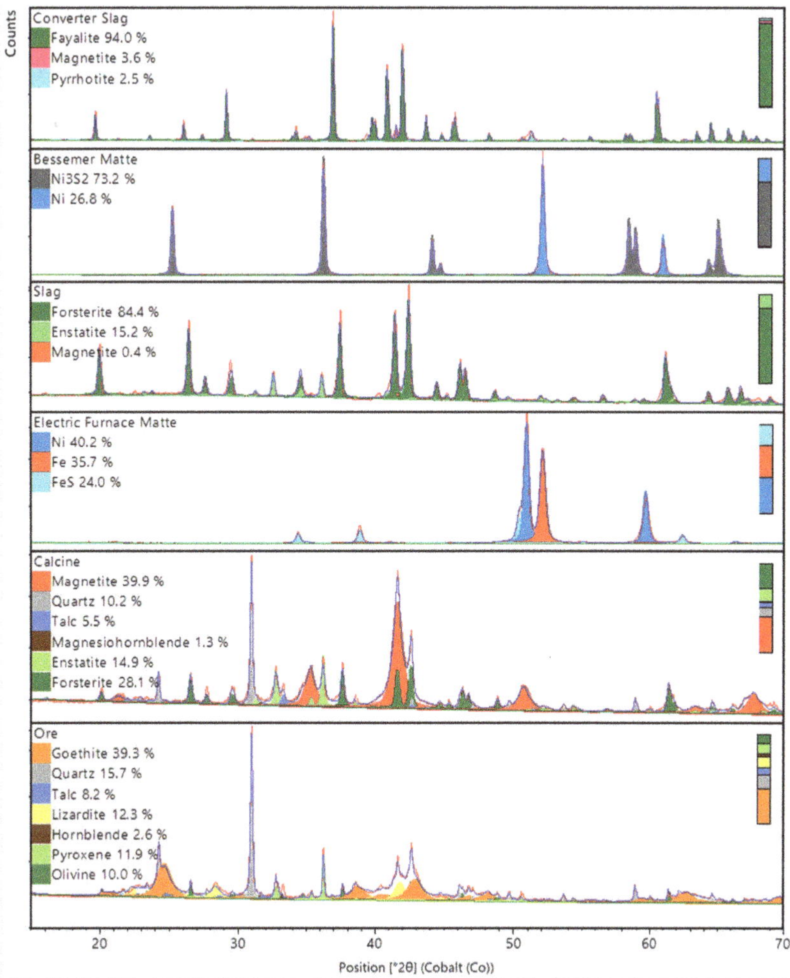

Figure 13. Example of a full-pattern Rietveld quantification for several processed materials of the pyrometallurgical extraction of nickel from laterites (Ore R_{wp} = 4.6, Calcine R_{wp} = 5.0, Electric Furnace Matte R_{wp} = 5.5, Slag R_{wp} = 5.5, Bessemer Matte R_{wp} = 2.2, Converter Slag R_{wp} = 6.3).

Table 4. Identified minerals in the process samples.

Mineral	Formula	References
Magnetite	Fe_3O_4	[44]
Hornblende	$(Ca,Na)_{2-3}(Mg,Fe,Al)_5(Al,Si)_8O_{22}(OH,F)_2$	[45]
Pyrrhotite	$Fe_{(1-x)}S$	[46]
Fayalite (Olivine)	Fe_2SiO_4	[47]
Nickel	Ni	[48]
Heazlewoodite	Ni_3S_2	[49]
Iron	Fe	[50]
Troilite	FeS	[51]

Lizardite is transformed to forsterite [52] in the calcination. The electric furnace matte consists of 40.2% nickel metal, 35.7% iron metal and 24% iron sulfide (troilite) whereas the corresponding slag consists mainly of silicates (forsterite and enstatite). The final product of the pyrometallurgical processing, Bessemer matte, has a nickel content of about 80%, consisting of 27% nickel metal and 73% nickel sulfide. The converter slag is rich in iron silicate and minor amounts of magnetite and pyrrhotite.

The examples demonstrate that frequent and fast mineralogical monitoring during the processing of nickel laterite ores can be a highly valuable tool to stabilize the furnace operation and increase the production of nickel matte despite lower feed grades.

4. Conclusions

X-ray diffraction (XRD) can be used as a fast and powerful tool to monitor nickel laterite mining and downstream processing. XRD is a fast and reliable technique to monitor the mineralogical composition of run-of-mine (RoM) ore, ore blends, calcine, matte and slag.

The mineralogy of the analyzed mine samples represents a typical nickel laterite profile with high amounts of goethite in the lateritic part and Mg-Fe-Silicates in the saprolitic part. The nickel concentration is correlated with the mineral Lizardite. Cluster analysis allows fast sorting of samples in groups of similar mineral composition and can be used as pass/fail analysis tool to sort nickel laterites. It allows objective grade control based on the mineralogical composition. Fuzzy clustering even allows the detection of mixtures of ores from different horizons or mineralogical domains.

Each step of the pyrometallurgical processing can be monitored for the phase composition of products (matte) and waste material (slag). This can lead to increased profitability of a mining operation. Table 5 summarizes the values that mineralogical monitoring using X-ray diffraction can deliver.

Table 5. Value and parameter to increase the profitability of nickel laterite processing.

Value	Tool
Optimization of ore blends from various nickel laterite deposits	Cluster analysis
Adjustment of superheat in the feed and optimization of energy costs	Mineralogy of ore blend
Control of mineralization acidity	Silicate composition
Prevention of highly corrosive slag causing erosion of the refractories	Silicate composition
Better reducibility in the furnace	Olivine content
Boost nickel recovery rates and reduction of metal loss in slag	Slag composition
Increase of cobalt recoveries.	Co-bearing minerals

Today's optics, detectors, and software for XRD analysis can provide rapid (<5 min) and accurate analyses, suitable for process control environments as well as research even with a small benchtop diffractometer. All evaluation methods such as cluster analysis, phase identification and quantification can run simultaneously and allow fast counteraction on changing conditions in the plant or in the mine. The complete analysis is ready for automation and can be easily included in existing automation lines.

Funding: This research received no external funding.

Data Availability Statement: All data were gathered and treated at Malvern Panalytical B.V laboratories in Almelo (The Netherlands).

Conflicts of Interest: The author declares no conflict of interest.

References

1. Butt, C.R.M.; Cluzel, D. Nickel Laterite Ore Deposits: Weathered Serpentinites. *Elements* **2013**, *9*, 123–128. [CrossRef]
2. Garvin, M.M.; Simon, M.J. A detailed assessment of global nickel resource trends and endowments. *Soc. Econ. Geol. Inc. Econ. Geol.* **2014**, *109*, 1813–1814.
3. DERA. *Batterierohstoffe für Die Elektromobilität*; DERA Themenheft: Berlin, Germany, 2021; p. 26.

4. Meshram, P.; Pandey, B.D. Advanced review on extraction of nickel from primary and secondary sources. *Miner. Process. Extr. Metall. Rev.* **2019**, *40*, 157–193. [CrossRef]

5. Agatzini-Leonardou, S.; Tsakiridis, P.E.; Oustadakis, P.; Karidakis, T.; Katsiapi, A. Hydrometallurgical process for the separation and recovery of nickel from sulphate heap leach liquor of nickelliferrous laterite ores. *Miner. Eng.* **2009**, *22*, 1181–1192. [CrossRef]

6. Tian, H.; Pan, J.; Zhu, D.; Yang, C.; Guo, Z.; Xue, Y. Imporved beneficiation of nickel and iron from a low-grade saprolite laterite by addition of limonitic laterite ore and $CaCO_3$. *J. Mater. Res. Technol.* **2020**, *9*, 2578–2589. [CrossRef]

7. Lv, X.W.; Bai, C.G.; He, S.P.; Huang, Q.Y. Mineral change of Philippine and Indonesian nickel laterite ore during sintering and mineralogy of the sinter. *ISIJ Int.* **2010**, *50*, 380–385. [CrossRef]

8. Brand, N.W.; Butt, C.R.M.; Elias, M. Nickel laterites: Classification and features. *AGSO J. Aust. Geol. Geophys.* **1998**, *17*, 81–88.

9. Maurizot, P.; Sevin, B.; Iseppi, M.; Giband, T. Nickel-Bearing Laterite Deposits in Accretionary Context and the Case of New Caledonia: From the Large-Scale Structure of Earth to Our Everyday Appliances. *GSA Today* **2019**, *29*, 4–10. [CrossRef]

10. Brindley, G.W.; Hang, P.T. The nature of garnierites—I. Structures, Chemical Composition and Color Characteristic. *Clays Clay Miner.* **1973**, *21*, 27–40. [CrossRef]

11. Brindley, G.W.; Maksimovic, Z. The nature and nomenclature of hydrous nickel-containing silicates. *Clay Miner.* **1974**, *10*, 271–277. [CrossRef]

12. Wells, M.A.; Ramanaidou, E.R.; Verrall, M.; Tessarolo, C. Mineralogy and crystal chemistry of "garnierites" in the Goro lateritic nickel deposit, New Caledonia. *Eur. J. Miner.* **2009**, *21*, 467–483. [CrossRef]

13. Horn, R.A.; Bacon, W.G. *Goro Nickel-Cobalt Project Located in French Overseas Territorial Community (Collectivite Territoriale) of New Caledonia*; Goro Nickel Technical Report; Goro Nickel: Noumea, New Caledonia, 2002; p. 116.

14. Degen, T.; Sadki, M.; Bron, E.; König, U.; Nénert, G. The HighScore suite. *Powder Diffr.* **2014**, *29*, S13–S18. [CrossRef]

15. COD Crystallography Open Database. Available online: http://www.crystallography.net/cod/ (accessed on 24 October 2021).

16. Rietveld, H.M. A profile refinement method for nuclear and magnetic structures. *J. Appl. Crystallogr.* **1969**, *2*, 65–71. [CrossRef]

17. Rietveld, H.M. Line profiles of neutron powder-diffraction peaks for structure refinement. *Acta Crystallogr.* **1967**, *22*, 151–152. [CrossRef]

18. Young, R.A. *The Rietveld Method. International Union of Crystallography*; Oxford University Press: Oxford, UK, 1993; p. 298.

19. Kelley, L.A.; Gardner, S.P.; Sutcliffe, M.J. An automated approach for clustering an ensemble of NMR-derived protein structures into conformationally related subfamilies. *Protein Eng. Des. Sel.* **1996**, *9*, 1063–1065. [CrossRef] [PubMed]

20. Liao, B.; Chen, J. The application of cluster analysis in X-ray diffraction phase analysis. *J. Appl. Crystallogr.* **1992**, *25*, 336–339. [CrossRef]

21. Van der Maaten, L.J.P.; Hinton, G.E. Visualizing High-Dimensional Data Using t-SNE. *J. Mach. Learn. Res.* **2008**, *9*, 2579–2605.

22. Sato, M.; Sato, Y.; Jain, L.C. *Fuzzy Clustering Models and Applications, Studies in Fuzziness and Soft Computing*; Springer: New York, NY, USA, 1997; Volume 9, p. 122.

23. Everitt, B.S.; Landau, S.; Leese, M. *Cluster Analysis*, 5th ed.; Wiley: London, UK, 2011; 346p.

24. Garnier, J. Essai sur la geologie et les ressources minerales de la Nouvelle-Caledonie. *Ann. Des Mines* **1867**, *6*, 1–92.

25. Faust, G.T. The hydrous nickel-magnesium silicgroup—The garnierite group. *Am. Min.* **1966**, *51*, 279–298.

26. Glasser, M.E. Note sur une espèce minérale nouvelle, la népouite, silicate hydraté de nickel et de magnésie. *Bull. De La Société Française De Minéralogie* **1907**, *30*, 17–28. [CrossRef]

27. Brindley, G.W.; Wan, H.M. Composition, structures, and thermal behavior of nickel-containing minerals in the lizardite-nepouite series. *Am. Mineral.* **1975**, *60*, 863–871.

28. Gamaletsos, P.N.; Kalatha, S.; Godelitsas, A.; Economou-Eliopoulos, M.; Göttlicher, J.; Steininger, R. Arsenic distribution and speciation in the bauxitic Fe-Ni-laterite ore deposit of the Patitira mine, Lokris area (Greece). *J. Geochem. Explor.* **2018**, *194*, 189–197. [CrossRef]

29. Samouhos, M.; Godelitsas, A.; Nomikou, C.; Taxiarchou, M.; Tsakiridis, P.; Zavasnik, J.; Gamaletsos, P.N.; Apostolikas, A. New insights into nanomineralogy and geochemistry of Ni-laterite ores from central Greece (Larymna and Evia depositis). *Geochemistry* **2018**, *12*, 5.

30. Alvarez, M.; Sileo, E.E.; Rueda, E.H. Structure and reactivity of synthetic Co-substituted goethites. *Am. Miner.* **2008**, *93*, 584–590. [CrossRef]

31. Gasser, U.G.; Jeanroy, E.; Mustin, C.; Barres, O.; Nüesch, R.; Berthelin, J.; Herbillon, A.J. Properties of synthetic goethites with Co for Fe substitution. *Clay Miner.* **1996**, *31*, 465–476. [CrossRef]

32. Cornell, R.M.; Schwertmann, U. *The Iron Oxides*; VHC Verlagsgesellschaft: Weinheim, Germany, 1996; pp. 35–47.

33. Cornell, R.M.; Giovanoli, R. Effect of cobalt on the formation of crystalline iron oxides from ferrihydrite in alkaline media. *Clays Clay Min.* **1989**, *37*, 65–70. [CrossRef]

34. Cornell, R.M. Simultaneous incooperation od Mn, Ni, Co in the goethite (α-FeOOH) structure. *Clay Miner.* **1991**, *2*, 427–430. [CrossRef]

35. Pozas, R.; Rojas, T.C.; Ocaña, M.; Serna, C.J. The Nature of Co in Synthetic Co-substituted Goethites. *Clays Clay Miner.* **2004**, *52*, 760–766. [CrossRef]

36. Blake, R.L.; Hessevick, R.E.; Zoltai, T.; Finger, L.W. Refinement of the hematite structure. *Am. Min.* **1966**, *51*, 123–129.

37. Saalfeld, H.; Wedde, M. Refinement of the crystal structure of gibbsite, $Al(OH)_3$. *Z. Für Krist.* **1974**, *139*, 129–135. [CrossRef]

38. Gualtieri, A. Accuracy of XRPD QPA using the combined Rietveld–RIR method. *J. Appl. Crystallogr.* **2000**, *33*, 267–278. [CrossRef]

39. Guggenheim, S.; Zhan, W. Effect of temperature on the structures of lizardite-1T and lizardite-2H1. *Can. Min.* **1998**, *36*, 1587–1594.
40. Perdikatsis, B.; Burzlaff, H. Strukturverfeinerung am Talk $Mg_3[(OH)_2Si_4O_{10}]$. *Z. Für Krist. Cryst. Mater.* **1981**, *156*, 177–186. [CrossRef]
41. Nestola, F.; Gatta, G.D.; Ballaran, T.B. The effect of Ca substitution on the elastic and structural behavior of orthoenstatite. *Am. Miner.* **2006**, *91*, 809–815. [CrossRef]
42. Ottonello, G.; Princivalle, F.; Della Giusta, A. Temperature, composition, and fO_2 effects on intersite distribution of Mg and Fe^{2+} in olivines. *Phys. Chem. Miner.* **1990**, *17*, 301–312. [CrossRef]
43. Bessemer, H. *Sir Henry Bessemer, F.R.S. An Autobiography*; Offices of "Engineering": London, UK, 1905; p. 176.
44. Nakagiri, N.; Manghnani, M.H.; Ming, L.C.; Kimura, S. Crystal structure of magnetite under pressure. *Phys. Chem. Miner.* **1986**, *13*, 238–244. [CrossRef]
45. Mancini, F.; Sillanpaa, R.; Marshall, B.; Papunen, H. Magnesian hornblende from a metamorphosed ultramafic body in southwestern Finland: Crystal chemistry and petrological implications. *Can. Mineral.* **1996**, *34*, 835–844.
46. Elliot, A.D. Structure of pyrrhotite 5C (Fe_9S_{10}). *Acta Crystallogr. Sect. B Struct. Sci.* **2010**, *66*, 271–279. [CrossRef]
47. Lottermoser, W.; Steiner, K.; Grodzicki, M.; Jiang, K.; Scharfetter, G.; Bats, J.W.; Redhammer, G.; Treutmann, W.; Hosoya, S.; Amthauer, G. The electric field gradient in synthetic fayalite α-Fe_2SiO_4 at moderate temperatures. *Phys. Chem. Miner.* **2002**, *29*, 112–121. [CrossRef]
48. Leineweber, A.; Jacobs, H.; Hull, S. Ordering of Nitrogen in Nickel Nitride Ni_3N Determined by Neutron Diffraction. *Inorg. Chem.* **2001**, *40*, 5818–5822. [CrossRef]
49. Fleet, M.E. The crystal structure of heazlewoodite, and metallic bonds in sulfide minerals. *Am. Min.* **1977**, *62*, 341–345.
50. Woodward, P.M.; Suard, E.; Karen, P. Structural Tuning of Charge, Orbital, and Spin Ordering in Double-Cell Perovskite Series between $NdBaFe_2O_5$ and $HoBaFe_2O_5$. *J. Am. Chem. Soc.* **2003**, *125*, 8889–8899. [CrossRef] [PubMed]
51. Bertaut, F. La structure de sulfure de fer. *J. De Phys. Et Du Radium* **1954**, *15*, 775.
52. Setiawan, I.; Febrina, E.; Subagja, R.; Harjanto, S.; Firdiyono, F. Investigations on mineralogical characteristics of Indonesian nickel laterite ores during the roasting process. *IOP Conf. Ser. Mater. Sci. Eng.* **2019**, *541*, 012038. [CrossRef]

 *minerals*

Article

Monitoring of Lithium Contents in Lithium Ores and Concentrate-Assessment Using X-ray Diffraction (XRD)

Herbert Pöllmann [1,*] and Uwe König [2]

[1] Department of Mineralogy, Institute of Geosciences and Geography,
Martin-Luther Halle-Wittenberg University, 06108 Halle, Germany
[2] Malvern Panalytical B.V., 7602 EA Almelo, The Netherlands; uwe.konig@panalytical.com
* Correspondence: herbert.poellmann@geo.uni-halle.de

Citation: Pöllmann, H.; König, U. Monitoring of Lithium Contents in Lithium Ores and Concentrate-Assessment Using X-ray Diffraction (XRD). *Minerals* **2021**, *11*, 1058. https://doi.org/10.3390/min11101058

Academic Editor: Fang Xia

Received: 22 July 2021
Accepted: 14 September 2021
Published: 28 September 2021

Abstract: Lithium plays an increasing role in battery applications, but is also used in ceramics and other chemical applications. Therefore, a higher demand can be expected for the coming years. Lithium occurs in nature mainly in different mineralizations but also in large salt lakes in dry areas. As lithium cannot normally be analyzed using XRF-techniques (XRF = X-ray Fluorescence), the element must be analyzed by time consuming wet chemical treatment techniques. This paper concentrates on XRD techniques for the quantitative analysis of lithium minerals and the resulting recalculation using additional statistical methods of the lithium contents. Many lithium containing ores and concentrates are rather simple in mineralogical composition and are often based on binary mineral assemblages. Using these compositions in binary and ternary mixtures of lithium minerals, such as spodumene, amblygonite, lepidolite, zinnwaldite, petalite and triphylite, a quantification of mineral content can be made. The recalculation of lithium content from quantitative mineralogical analysis leads to a fast and reliable lithium determination in the ores and concentrates. The techniques used for the characterization were quantitative mineralogy by the Rietveld method for determining the quantitative mineral compositions and statistical calculations using additional methods such as partial least square regression (PLSR) and cluster analysis methods to predict additional parameters, like quality, of the samples. The statistical calculations and calibration techniques makes it especially possible to quantify reliable and fast. Samples and concentrates from different lithium deposits and occurrences around the world were used for these investigations. Using the proposed XRD method, detection limits of less than 1% of mineral and, therefore down to 0.1% lithium oxide, can be reached. Case studies from a hard rock lithium deposit will demonstrate the value of mineralogical monitoring during mining and the different processing steps. Additional, more complex considerations for the analysis of lithium samples from salt lake brines are included and will be discussed.

Keywords: lithium; quantification; XRD; PLSR; clustering; Rietveld; cluster analysis; spodumene; petalite; lepidolite; triphylite; zinnwaldite; amblygonite

1. Introduction

Lithium is an element in the chemical periodic system with ordinal number 3. The formation and details on the metal are summarized by [1–5]. Almost half of the lithium production is nowadays increasingly used in the fabrication of batteries [5,6]. Therefore, an increasing demand in the metal supply is visible [7,8]. Other uses of lithium are in ceramics industry, grease, polymers and air treatment among others. The main lithium resources are coming from lithium salt lake brines in arid areas and from different lithium containing minerals, often concentrated in economic mining sites [9–13]. Lithium resources in Europe were summarized by [14]. The worldwide situation was described by [15–17]. Some lithium containing clays are also promising nowadays. Lithium forms 124 mineral species [18,19]. 44% of all lithium minerals occur in LCT-pegmatites and associated metasomatic rocks [20,21]. Other main sources of lithium minerals are non-LCT pegmatites and their metasomatic rocks, metasomatic rocks not associated with pegmatites, and manganese deposits [19]. A list of widely abundant minerals is given in [18]. However, not all of these are enriched in quantities to be mined. The mineral with the highest lithium-content is the rare Zabuyelite Li_2CO_3 (18.79%). A summary of lithium demand and supply is given in [22]. By analyzing elements associated in LCT pegmatites (LCT = lithium-cesium tantalum pegmatites) [23] (Ga, Rb, Nb, Sn, Cs, Ta, Tl) satisfactorily lithium can be predicted [24] by μ-XRF. Another basis for lithium is also the occurrence in salt lakes in dry areas. The lithium must be concentrated by thermal and leaching processes [25–30]. A review of enrichment techniques is given by [31]. Some investigations are included in this publication. Li-Minerals, which can be mined in large quantities, as a basic source from geological enrichment with their chemical compositions and XRD data files, are summarized in Table 1. The lithium determination in these mineral concentrates will be described.

Table 1. Important lithium-bearing minerals, compositions and crystal structure files.

Mineral	Composition	Lithium-Content in wt.%	Lithium Oxide Content in wt.%	ICSD-No.	Literature
Spodumene	$LiAl[Si_2O_6]$—(α, ß, γ)	3.73	8.03	30,521 9668 (α) 14,235 (ß) 69,221 (γ) Virgilite	[32]
Petalite	$LiAl[Si_4O_{10}]$	3.09	4.50	100,348	[33]
Lepidolite Polylithionite	$KLi_2[AlSi_3O_{10}/(OH,F)_2]$	3.58 3.00	7.7 6.46	30,785 34,336 (with F)	[30]
Zinnwaldite	$KLiFeAl[AlSi_3O_{10}/(F,OH)_2]$		3.42	432,226	[34]
Amblygonite/ Montebrasite	$LiAl[PO_4F]$/ $LiAl[PO_4OH]$	3.44/3.80 4.74	10.21 7.49 (at 5% Na_2O)	26,513 68,925 (OH/F) 68,921 (OH)	[18]
Lithiophilite/ Triphylite	$LiMn[PO_4]$/ $LiFe[PO_4]$	4.43 4.40	9.53 9.47	75,283 72,545	[35]

For these lithium minerals from definite occurrences, different calibration curves were set up and statistical methods were successfully introduced to determine the relevant lithium oxide content of these mixtures. The lithium content of these Li-minerals increases from Zinnwaldite, Petalite, Lepidolite, Spodumene, and Triphylite to Amblygonite/Montebrasite. It must also be taken into account that these minerals can form solid solutions which do influence these absolute lithium contents and must be adapted and determined separately [36]. In Figure 1 the theoretical calibration curves of the six lithium containing minerals, with ideal composition, showing their maximum lithium contents, are given. In practical work these curves with the lithium contents of the respective occurrence must be adapted due to the investigated mineral composition and ore compositions from different geological occurrences.

The used lithium minerals in this study come from different origins. These determined compositions can be representative for other occurrences, but must otherwise simply be adapted, mainly when different solid solutions of these minerals do occur. For all calibrations, the standard mineral compositions of the relevant occurrences were used. Calibrations must be changed when larger differences in mineral compositions occur. The used lithium ores and their matrices can be seen in Figure 2a–f showing the different lithium minerals coming from varying origins. These used ores and concentrates are rather simple in mineralogical compositions as mineral paragenesis for these processed ores are very uniform. Many of these ores are only composed of additional quartz, and some also contain feldspar.

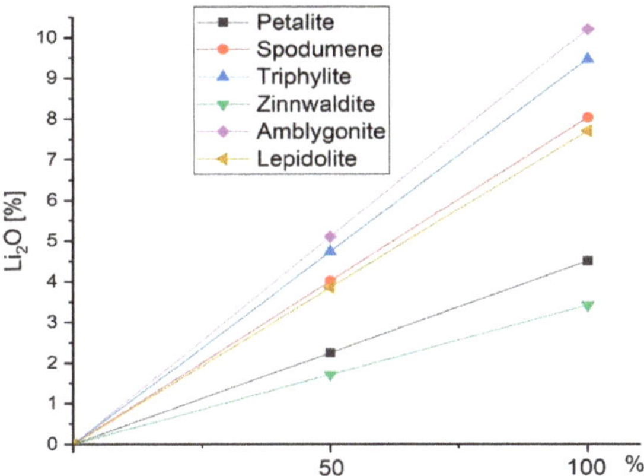

Figure 1. Maximum lithium oxide contents in different lithium minerals. Variation of lithium depending from content of the minerals (wt.%).

Figure 2. (**a**) Underground image of slightly green spodumene crystals (green crystal size ca. 1 m) in quartz matrix—MG—Brazil, (**b**) Montebrasite/Amblygonite occurrence with quartz—Serra Branca, Brazil, image: 30 m, (**c**) Triphylite with small quartz intergrowth—Hagendorf, Germany, image: 10 cm, (**d**) Zinnwaldite mica—Ore mountains/Germany, image: 5 cm, (**e**): Lepidolite in quartz/feldspar assemblage—Mangualde, Portugal, image: 10 cm, (**f**) Petalite/Rubicon mine–Namibia, image: 10 cm.

The lithium contents of the minerals must be carefully calculated due to their chemical compositions. Spodumene, petalite and triphylite minerals are easy to handle since their solid solutions are rather restricted. For Montebrasite/Amblygonite, Zinnwaldite and Lepidolite, the lithium contents must be determined in advance, since different solid solutions of these minerals may occur in different occurrences.

For all determinations it is also necessary to verify these contents from time to time, as some change may also be possible in the geological surroundings within one occurrence.

2. Experimental

A technical approach for lithium determination by quantitative mineral determination and following Li_2O calculation was performed using XRD powder techniques and different interpretation methods. Due to the non-possibility of a direct XRF analysis for lithium, a new method based on quantitative X-ray Diffraction of the lithium minerals, combined with calculation of lithium from their mineralogical composition, is proposed. Time-consuming wet chemical analysis procedure is only used from time to time for verification purposes. The following XRD techniques and combined calculating statistical procedures were applied for these quantitative mineral and lithium determinations:

2.1. Qualitative and Quantitative Mineral Analysis by XRD Combined with Rietveld Refinement Calculations

The XRD analysis was performed using a PANalytical Expert system (Panalytical Highscore Version 4.9, Almelo, The Netherlands) with X'celerator detector and program Highscore Plus for analytical treatment of qualitative mineral compositions. Mineral determinations were done by using the actual ICDD (PDF4, 2021) (International Centre for Diffraction Data, Newtown Square, PA, USA) database. Also, the same program was used for Rietveld [22,37] refinement and statistical procedures. The relevant ICSD (Inorganic Crystal Structure Database, FIZ Karlsruhe) structure files are summarized in Table 1. The sample preparation for XRD was conducted by filling fine milled powders into steel ring sample holders using the backload preparation methodology.

The following parameters for X-ray experiments were used (Table 2).

Table 2. XRD measurement parameters. (LFF = long fine focus spot).

Parameter	LFF-Tube-Cu	LFF-Tube-Co
Measurement range (2 theta)	5–70	5–90
Step (2 theta)	0.02	0.02
Counting time in s	10	10
Antiscatter slit	1/8	1/8
Soller slits	2.3	2.3
Voltage in kV	45	40
Tension in mA	30	35
ß-filter	Ni	Fe

Quantitative analysis of the samples was mainly made using Rietveld analysis in Highscore Plus program. Quantification of amorphous contents was made by the addition of an Inner standard (Rutile) of 10%. Background calculation was added as determined manually. Refinement control of the samples was performed using the Pseudo-Voigt profile function, scale factor, zero shift, unit cell and W-profile parameter. The verification of the determined contents was obtained by addition of definite amounts of mineral phases and following calculations and construction of the regression curves.

2.2. PLSR—Partial Least Squares Refinement Techniques for Different Mineral Mixtures

Full pattern Rietveld quantification [2,8,38,39] using crystal structures can be replaced by partial least squares refinement calculations using high score plus platform and the specific part of the program for correlating mineral contents and XRD results. The PLSR can be mainly used also in industrial applications to correlate mineral components with their quantitative contents. With this method no pure phases, crystal structures or modelling of peak shapes are necessary. For setting up this method a number of samples for calibration and validation are needed. Data are afterwards processed by calculating diagrams showing regression plots of reference values (X-axis) versus predicted values (Y-axis). Different scaling mode methods can be used to find the best calibration model (center method (the mean is subtracted from each value for the definite variable) and standardize method (the variables are standardized by subtracting the mean and dividing by the standard deviation). After calibration these models can be used for unknown samples to predict the defined property. With the used equipment, a precise result can be obtained within minutes and therefore the method can be used for process control. No specific preferred orientation refinement model was used, as this can influence the content of the minerals. Instead, for the calibration the preparation was standardized.

2.3. Cluster-Analysis of Obtained XRD Patterns of Mixtures

Cluster analysis is a statistical method [2,40–42] which can be used for obtaining rapid results in ore assessment by finding similar groups (clusters) which are more similar to each other than to those of other groups. The technique is mainly used for data reduction. It is applied here to find clusters containing similar grades of lithium contents in the ores (high, intermediate, and low grade) to obtain arguments for further ore treatment. Within this method additional possibilities can be used for further optimizations (PCA (principal component analysis), dendrogram, KGS (Kelley, Gardner, Sutcliffe) test and others). The dendrogram is a possibility to define different clusters using the cut off between less similar XRD patterns. Mainly, cluster analysis is used for data reduction of complex systems.

2.4. Principal Component Analysis—PCA

PCA plots [40,43] using the resulting eigenvectors, can create a three-dimensional arrangement of first three principal components and can show thereof the XRD data belonging to the different clusters.

2.5. Determination of Detection Limits for Different Lithium Minerals in Matrices

The minimal detection of minerals in the relevant matrices is the basic fact that must be done to see what is the lowest lithium concentration which can be detected. Despite these low concentrations not being as interesting as lithium ores, these mixtures can be classified in the different group of low content lithium ore or otherwise a different fourth grade can be added, meaning low lithium contents without any interest for lithium extraction. The lowest lithium concentration which can be measured is dependent on the possibility of identifying the lithium mineral in the matrix and is given for any lithium mineral used in this study.

Furthermore, the program High Score Plus was used for Rietveld refinement and some additional statistical procedures. The relevant ICSD structure files necessary for Rietveld analysis are summarized in Table 1.

For these investigations of lithium content determinations in lithium concentrates different localities with several relevant lithium minerals from deposits in Germany, Portugal, Namibia, Brazil, Finland, and Australia were used. The results for these 6 main important lithium minerals, ores and concentrates are presented. Different additional interesting results were obtained for complex analyses of multi-mineral lithium brine investigations originating from Chile.

3. Results for the Quantifications of Different Lithium Minerals in Binary Mixtures with Quartz

3.1. Quantification of Petalite—$LiAlSi_4O_{10}$-Quartz SiO_2

Petalite ore samples from the Namibian occurrence were used in a typical binary mixture with quartz and 10% of weight differences were used. These samples were used to determine the mineral contents and thereof the lithium contents. A calibration curve for Petalite [33] with different amounts (10% weight portions) of accompanying quartz was set up using XRD's with different contents of the minerals (Figure 3). It can easily be seen, that high and medium grade ores (contents of 100 weight% to 20 weight% Petalite) show strong Petalite peaks in XRD patterns, which can easily be identified.

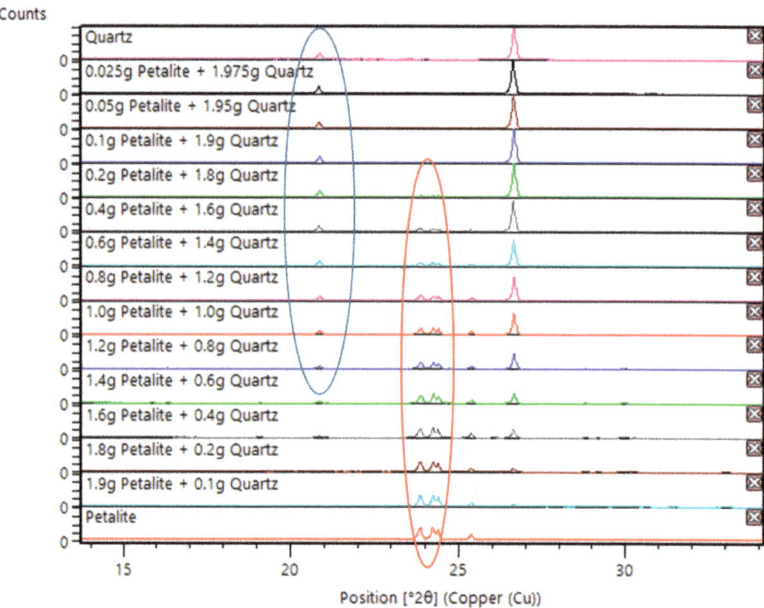

Figure 3. XRD patterns used for the calibration of petalite–quartz weight% mixtures (triplet of petalite peaks and main peak of quartz highlighted).

In the mixtures also the minimum content of the lithium mineral petalite, which can be detected, was determined. In Figure 4 the XRD patterns of pure petalite and a mixture of quartz with 5% of petalite is shown. Figure 4b shows several XRD patterns of binary mixtures and also the binary mixture at the detection limit of Petalite in a quartz matrix. It can be seen, that the main peak of petalite is easily detectable, meaning, that lithium contents down to 0.5% Li_2O can be detected. In Figure 5 the linear calibration curve using PLSR curves of petalite in quartz matrix is given.

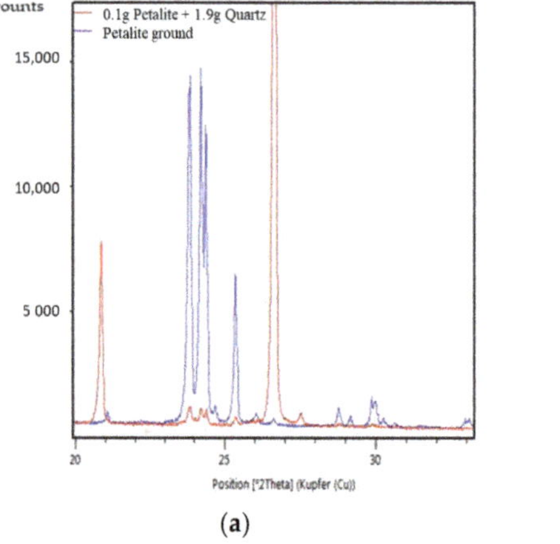

(**a**)

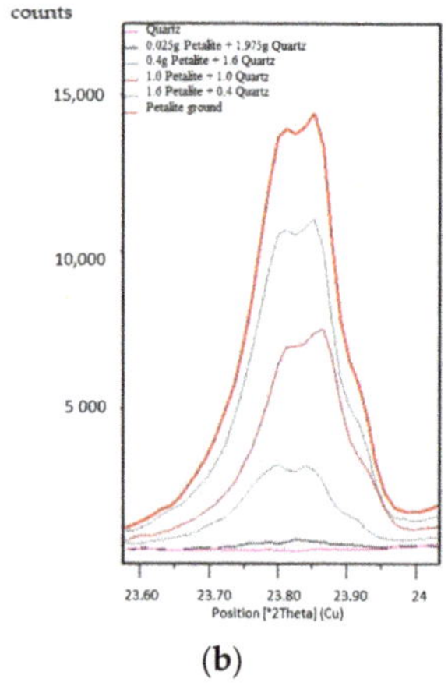

(**b**)

Figure 4. Determination limit of petalite in quartz matrix using main peaks of petalite. (**a**) Part of XRD pattern petalite and mixture of 95% quartz/5% Petalite; (**b**) Main peak of petalite in quartz matrices of different contents.

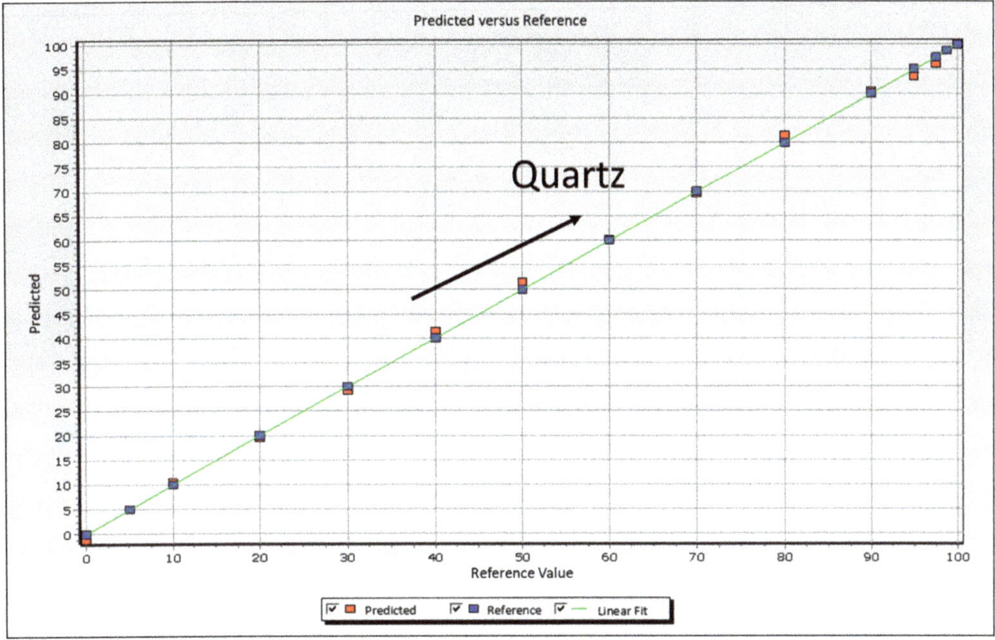

Figure 5. PLSR calibration curve for petalite–quartz mixtures.

From these different contents of petalite in lithium mineral quartz matrix the detection limit for lithium oxide can be derived (Table 3) showing the contents of lithium oxide derived from XRD patterns in Figure 4b. The PLSR calibration curve for petalite–quartz mixtures is given in Figure 5.

Table 3. Lithium oxide content as given from several selected petalite/quartz mixtures.

XRD-Mixture of Petalite/Quartz in %	Li$_2$O Content in %
Petalite 100	4.5
Petalite/Quartz 80/20	3.6
Petalite/Quartz 50/50	2.25
Petalite/Quartz 20/80	0.9
Petalite/Quartz 1.25/98.75	0.056

The cluster analysis brings similar contents of minerals into selective groups when necessary. Segments of three different clusters, containing areas with comparable lithium contents (high-medium-low grade ores) were determined in this case (Figure 6), but it is also possible to calculate other different clusters (and derived lithium contents thereof) in application cases when necessary. Here it is possible to determine for instance cut off values of economic lithium mineral contents.

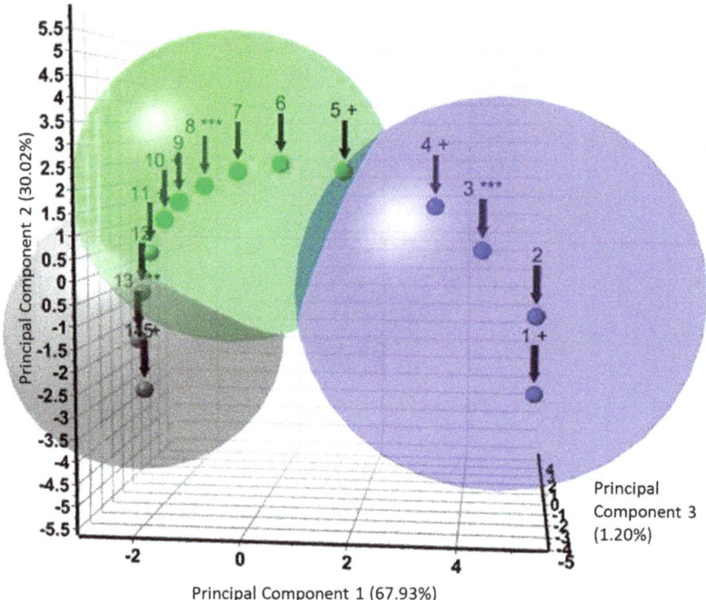

Figure 6. Clustering of petalite–quartz mixtures (three clusters with differing lithium contents). *** most representative sample of cluster, + are the most different ones in the clusters.

3.2. Quantification of Spodumene LiAlSi$_2$O$_6$-Quartz SiO$_2$

As spodumene is an important lithium mineral [32,44] some determinations on rather pure ores and concentrates were made. The results of the XRD calibration measurement of spodumene and quartz mixtures from Brazilian occurrence are summarized in Figure 7. The main peaks of spodumene and quartz are highlighted, to verify, that precise identifications of minerals can be made rather easily. The PLSR calibration curve for mixtures of spodumene and quartz is shown in Figure 8.

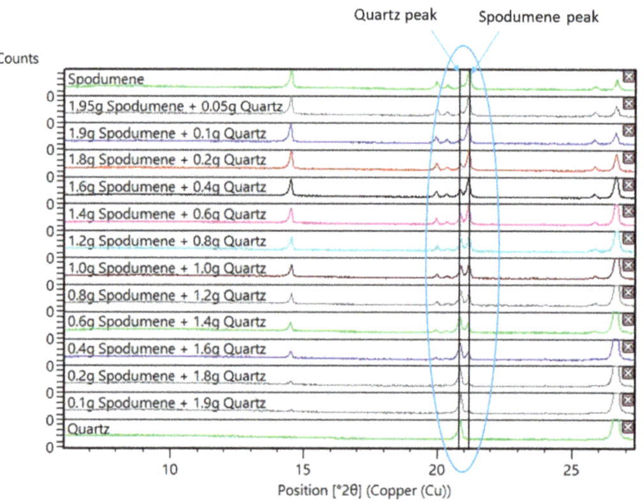

Figure 7. XRD patterns of different spodumene–quartz mixtures (main peak of quartz and spodumene highlighted).

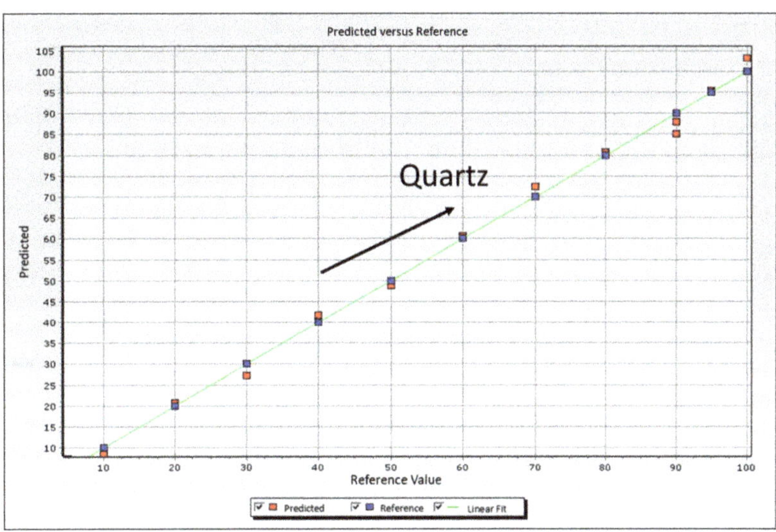

Figure 8. Calibration curve for spodumene–quartz mixtures.

As an example for Rietveld refinement control of the different mixtures, the Rietveld plot for a mixture of 95% spodumene and 5% of quartz and the calculated composition is given in Figure 9.

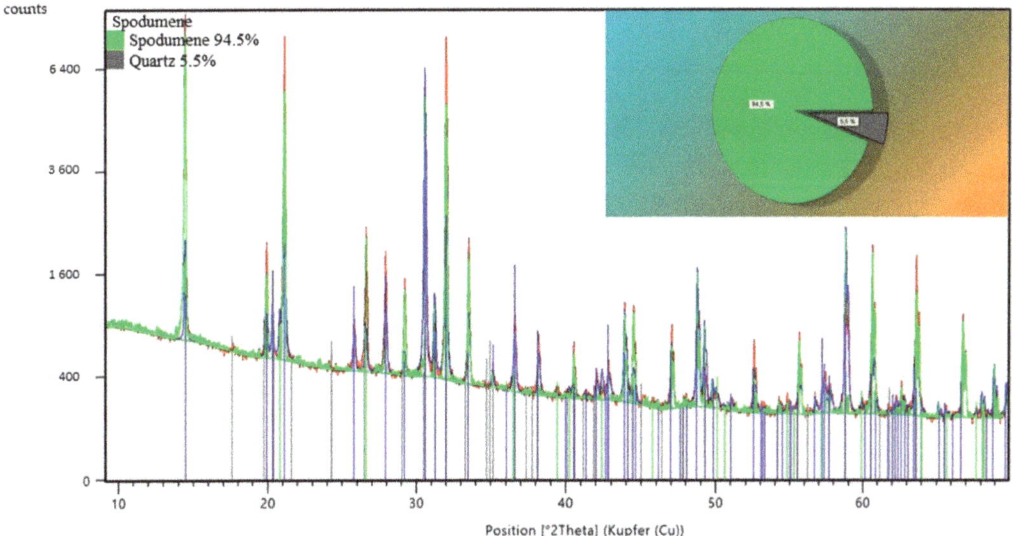

Figure 9. Rietveld refinement of a spodumene–quartz mixture.

It is also possible to cluster these spodumene containing mixtures in three different lithium grade ores as given in Figure 10 (low, medium, and one high lithium containing grade). The detection limit of spodumene was also determined.

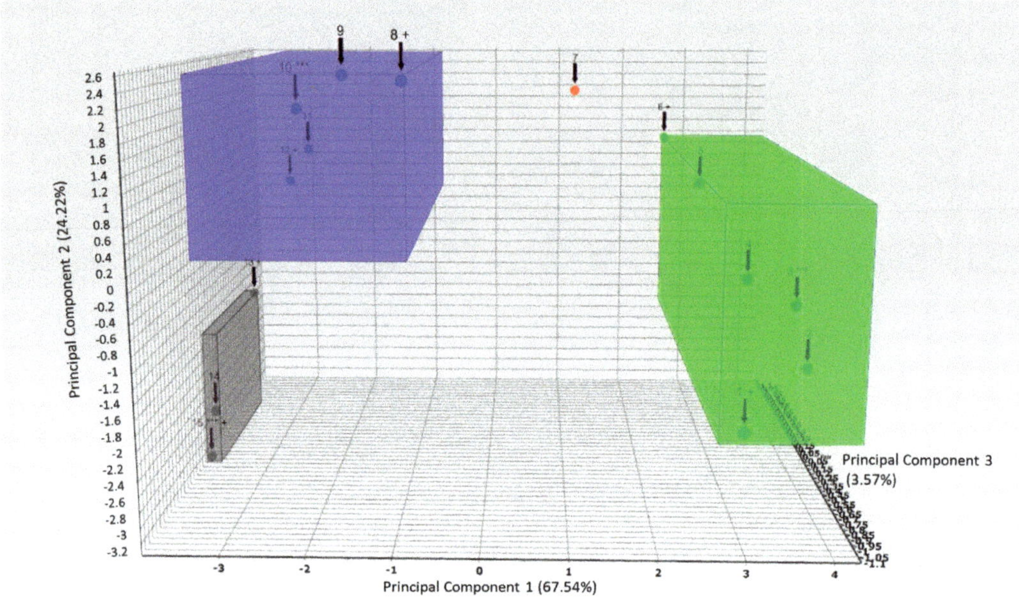

Figure 10. Clustering of spodumene–quartz mixtures (three clusters with different lithium contents). *** most representative sample of cluster, + are the most different ones in the clusters.

3.3. Quantification of Triphylite LiFePO$_4$-Quartz SiO$_2$

Investigations using Triphylite (LiFePO$_4$) (Hagendorf/Bavaria) [36] ore must primarily deal with the possible solid solution with Lithiophorite (LiMnPO$_4$). This composition for the geological occurrence is definitely used for making the calibration curves. Figure 11 shows the PLSR calibration curve for triphylite–quartz mixtures for a typical Hagendorf lithium ore.

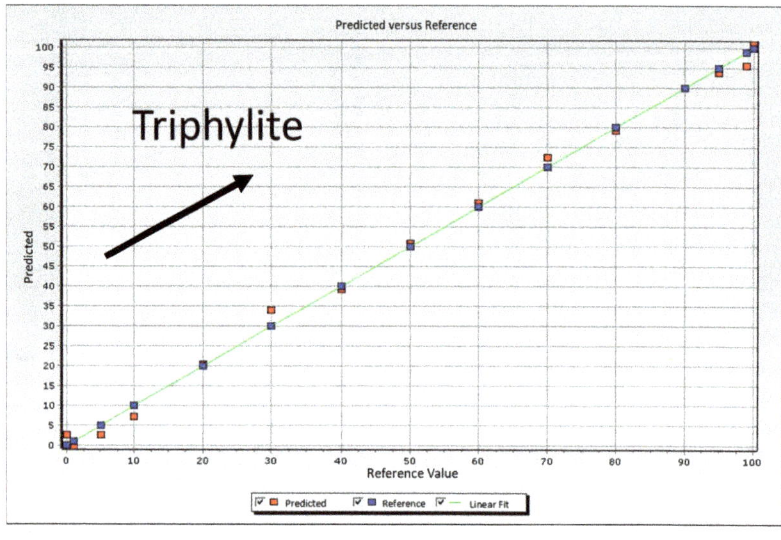

Figure 11. Calibration curve for triphylite–quartz mixtures.

The different mixtures were additionally clustered into four different triphylite containing lithium mineral mixtures (Figure 12). The detection limit for triphylite in quartz mixtures and the lithium content of these ores was determined as well (Figure 15).

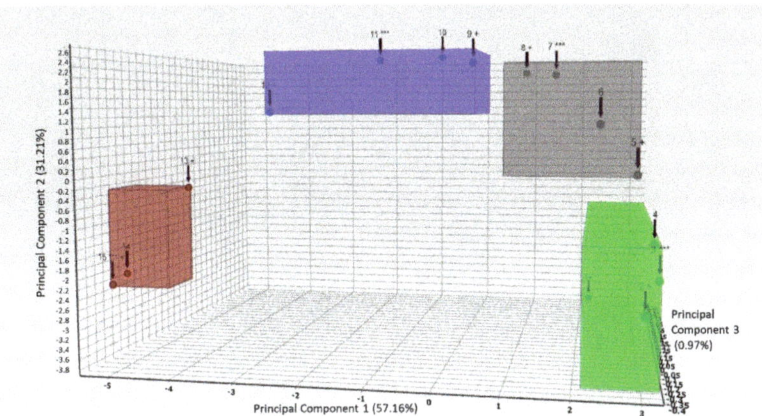

Figure 12. Clustering of triphylite–quartz mixtures showing four different lithium ore grades. *** most representative Scheme 2. contents, low triphylite, medium triphylite, and high triphylite contents. The XRD patterns of these mixtures and their calculated groups, including their similarities, are given in Figure 14. To show the relevant main peaks of the mineral quartz in these mixtures, with special emphasis to low amounts, Figure 15 is added.

The graphical way to determine detection limits of triphylite is given in Figure 13 by showing the relevant XRD patterns of quartz–triphylite mixtures. 1% of triphylite in a quartz matrix can easily be determined in these mineral mixtures.

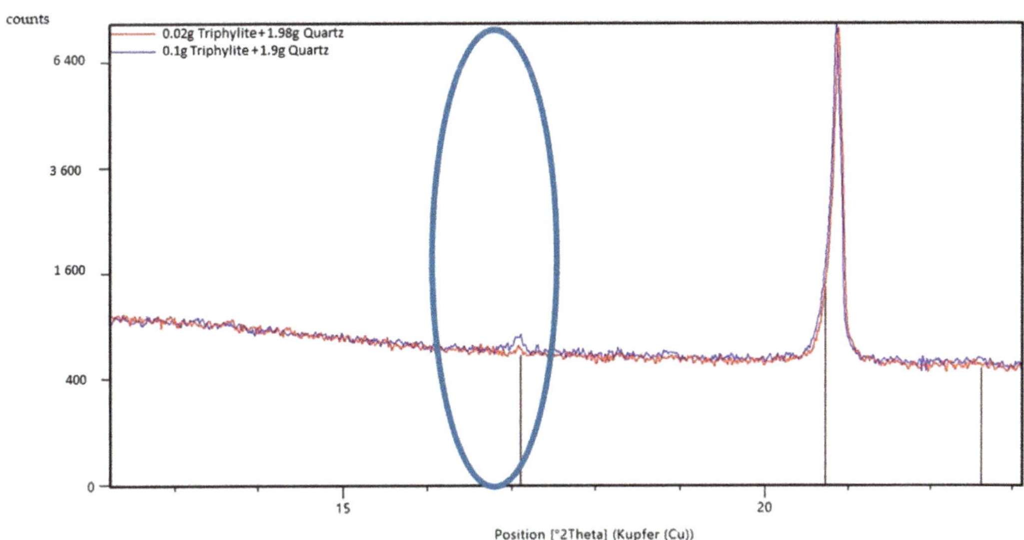

Figure 13. XRD pattern of small additions of triphylite to quartz matrix (1% Triphylite/99% Quartz).

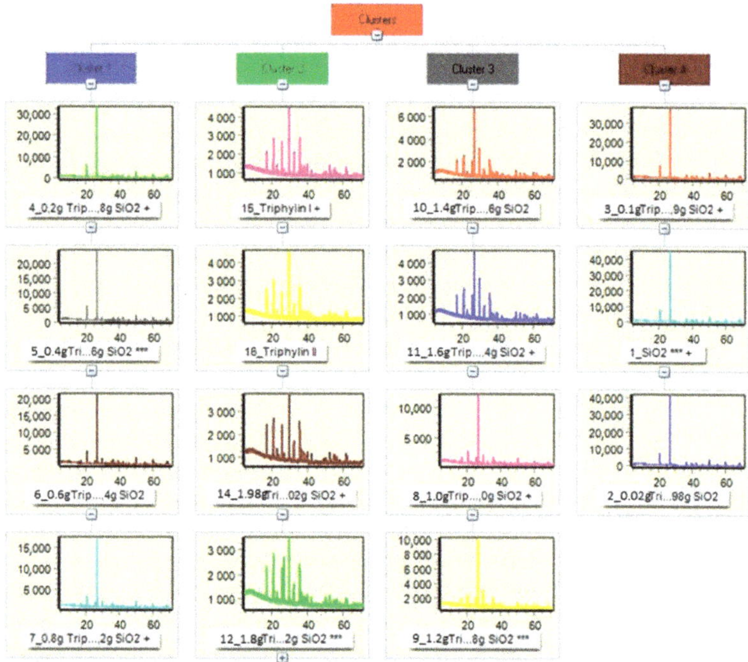

Figure 14. XRD patterns belonging to the respected determined clusters with different triphylite–quartz contents.

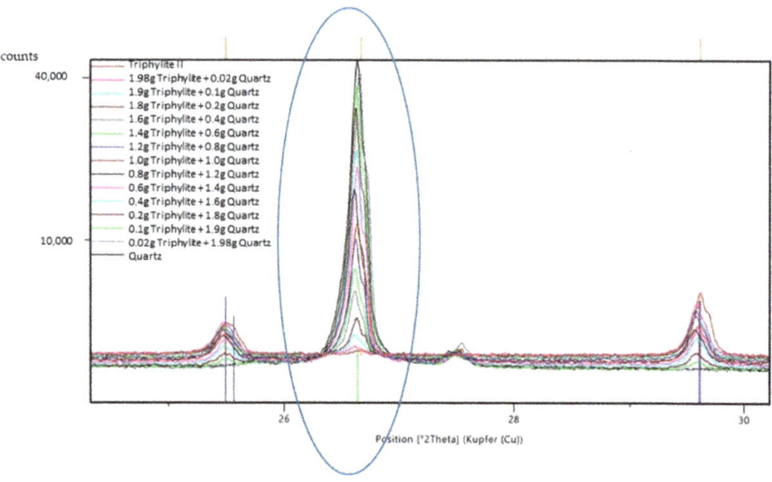

Figure 15. Main peak of quartz in triphylite–quartz calibration mixtures.

3.4. Quantification of Montebrasite LiAlPO$_4$(OH,F)-Quartz SiO$_2$

For the determination of different montebrasite (F-containing) contents from NE-Brazil in the mixtures with quartz a calibration curve using PLSR was set up. The derivation of this method must also include the solid solution of the minerals montebrasite and Amblygonite [45]. The XRD patterns of montebrasite/quartz mixtures are shown in Figure 16 and the PLSR calibration curve for montebrasite is given in Figure 17.

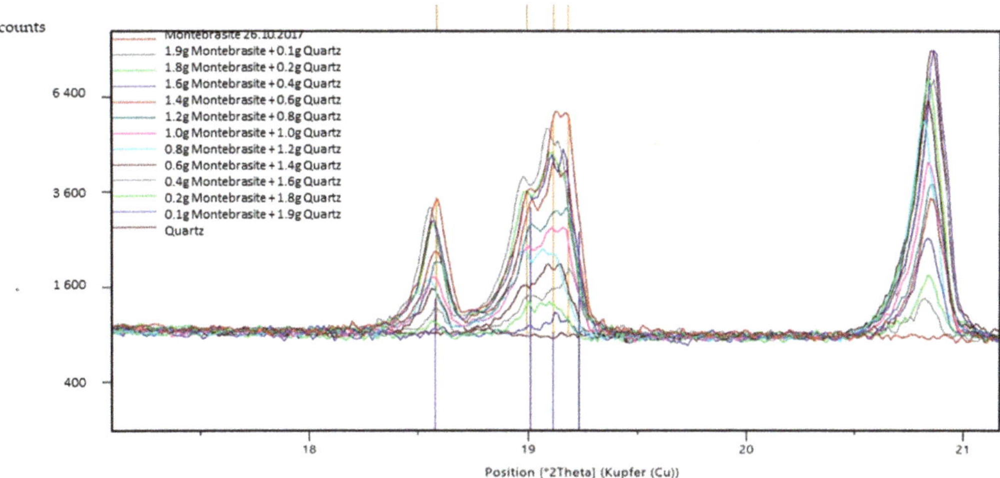

Figure 16. XRD patterns showing different contents of montebrasite–quartz mixtures.

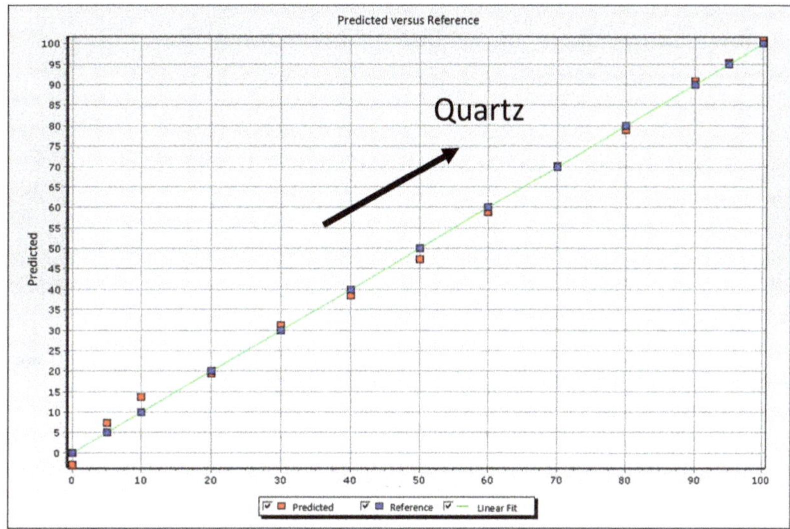

Figure 17. Calibration curve for montebrasite–quartz mixtures.

The clustering of XRD scans revealed four clusters of (a) SiO$_2$-rich, (b) Montebrasite-rich, (c) intermediate Montebrasite, and (d) intermediate SiO$_2$-contents (Figure 18).

The following cluster details showed the different XRD patterns which compose these separated four clusters with different montebrasite contents (high, medium, medium low, low) (Figure 19).

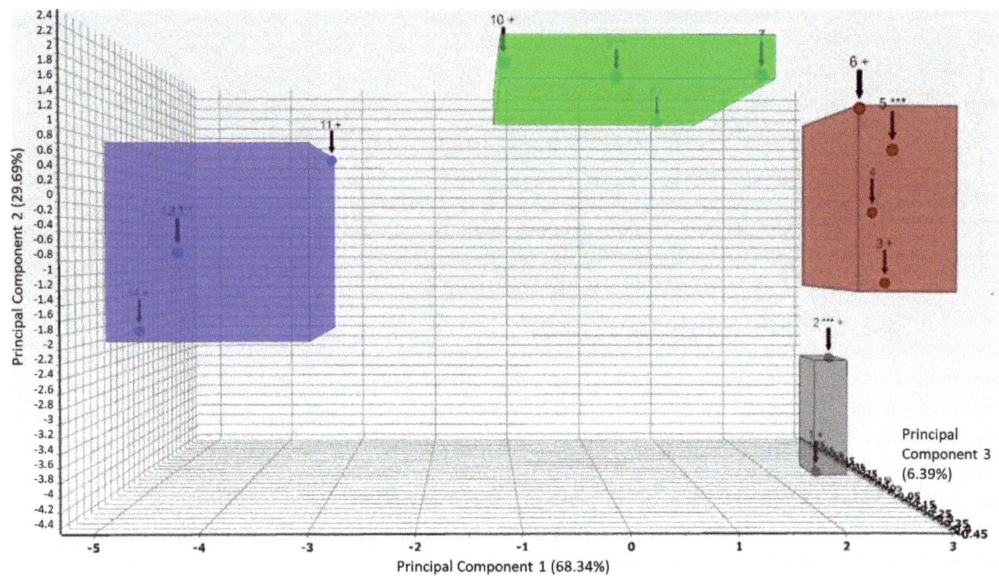

Figure 18. Clustering of montebrasite–quartz XRD diagrams. *** most representative sample of cluster, + are the most different ones in the clusters.

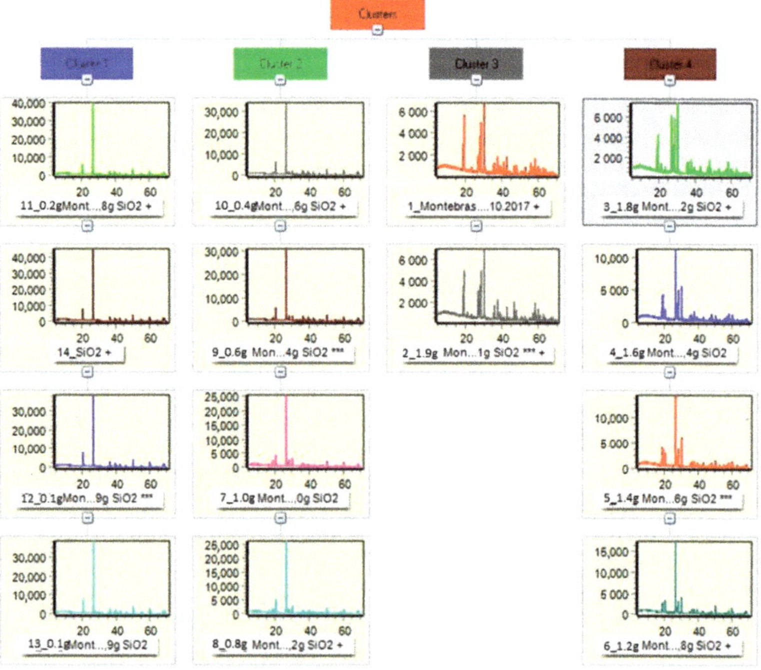

Figure 19. XRD patterns of quartz–montebrasite mixtures forming the four relevant calculated clusters.

3.5. Quantification of Lepidolite $KLi_2AlSi_3O_{10}(OH,F)_2$-Quartz SiO_2

For the calibration of lepidolite mica and quartz some samples from Portugal were used. For micas the preparation must be done very carefully to obtain a comparable kind of preferential orientation in all samples, as this can influence the intensity of the peaks. Especially for small amounts of mica, a highly preferential orientation can be included to increase the determination limit. The intensity of a main peak of lepidolite in the mixtures with quartz is given in Figure 20. The calculated PLSR calibration curve for lepidolite is given in Figure 21.

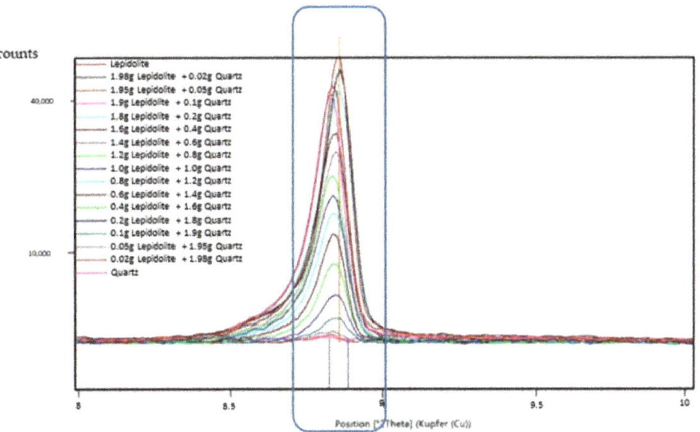

Figure 20. Details of XRD patterns showing main peak of lepidolite in different contents.

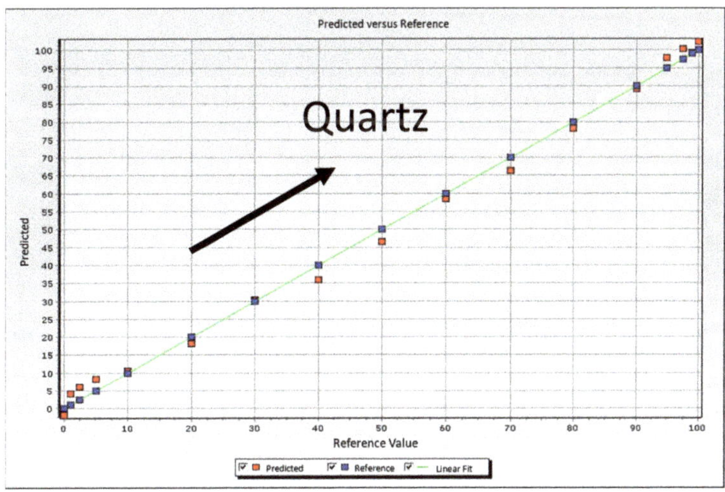

Figure 21. Calibration curve for lepidolite–quartz mixtures.

The results of the clustering into four lepidolite containing clusters and the relevant XRD patterns are given in Figures 22 and 23.

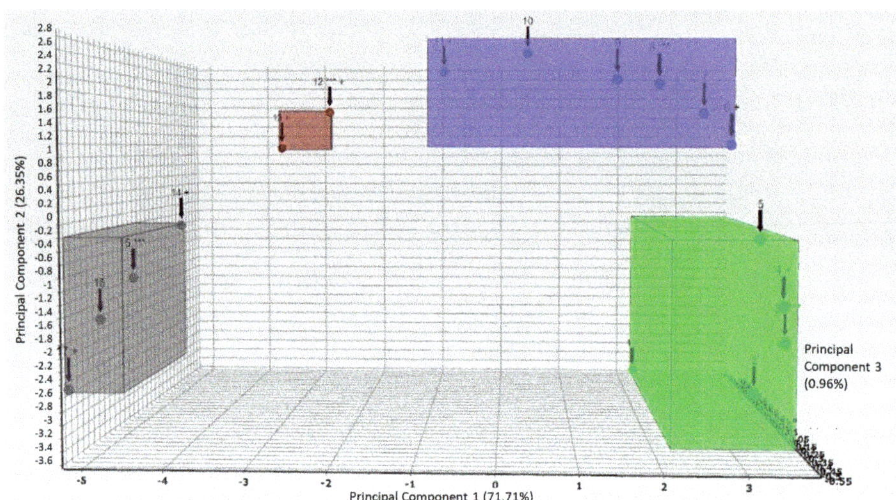

Figure 22. Clustering of XRD patterns of lepidolite–quartz mixtures. *** most representative sample of cluster, + are the most different ones in the clusters.

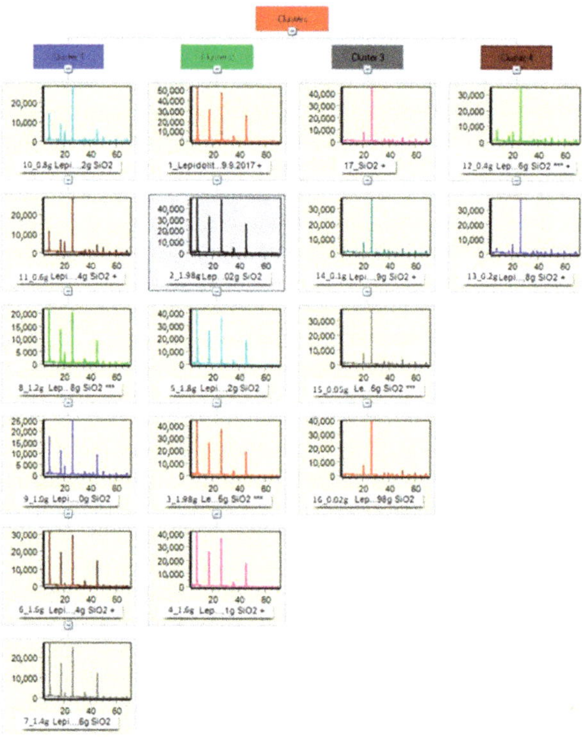

Figure 23. XRD patterns of quartz–lepidolite mixtures forming the relevant calculated clusters.

3.6. Calibration of Zinnwaldite $KLiFeAl_2Si_3O_{10}(F,OH)_2$-Quartz SiO_2

Using zinnwaldite [35] as a lithium source the same precautions in the preparation of XRD patterns as in lepidolite case for micas must be used controlling the preferential orientation of mica platelets. Typical XRD patterns showing varying contents of quartz and zinnwaldite peak areas are given in Figure 24. The calculated PLSR calibration curve for zinnwaldite-quartz mixtures is given in Figure 25.

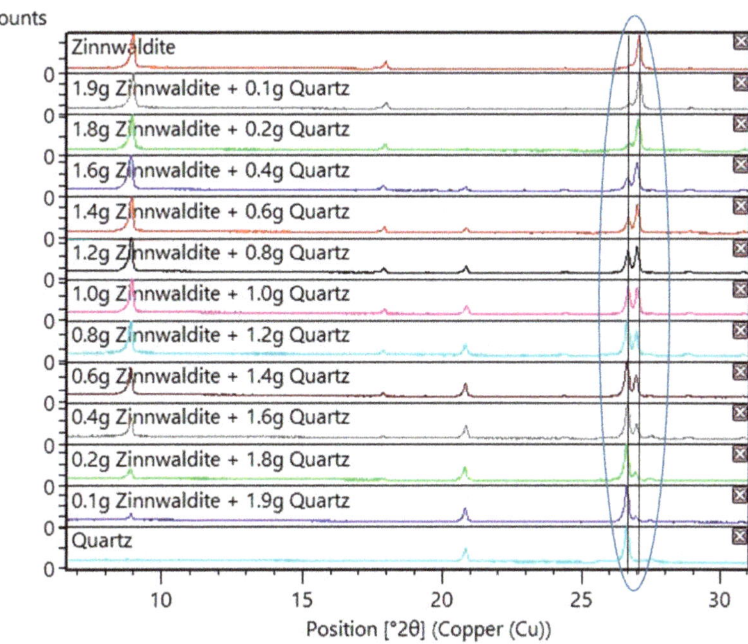

Figure 24. XRD patterns of different zinnwaldite–quartz mixtures with main peaks highlighted.

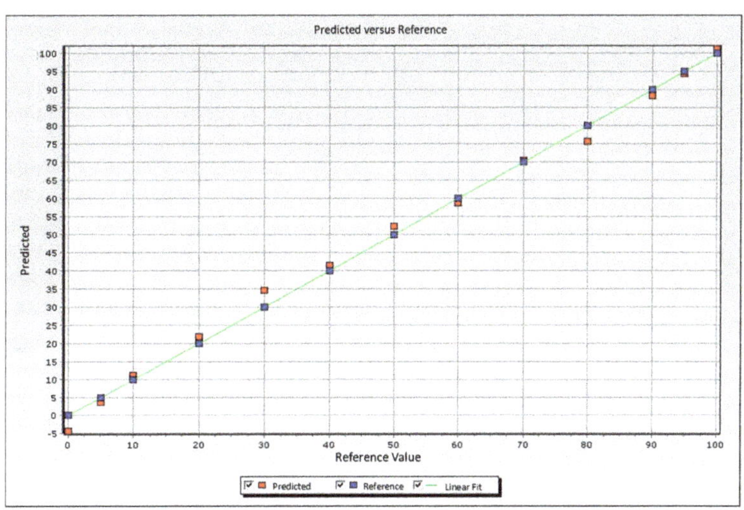

Figure 25. PLSR calibration curve for zinnwaldite–quartz mixtures.

Using the cluster calculation, the following four clusters containing different lithium contents based on zinnwaldite can be generated (Figure 26).

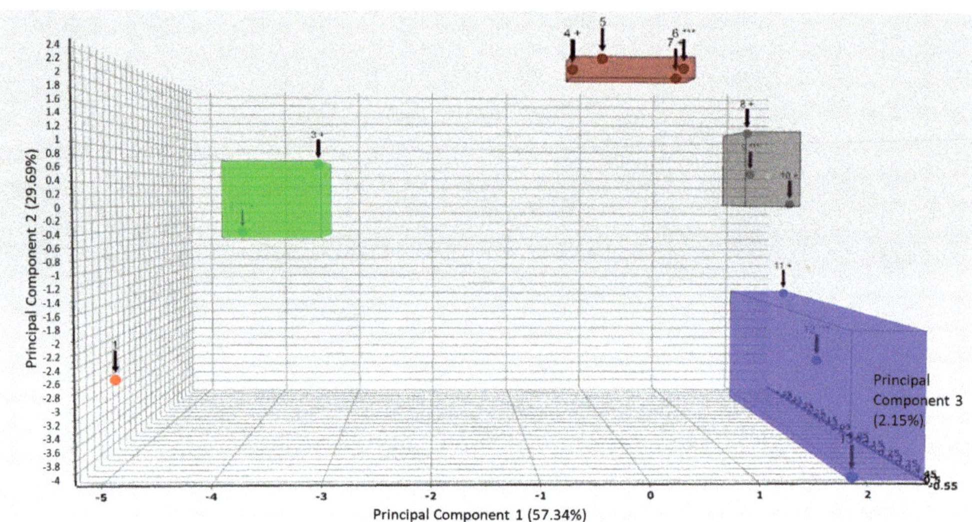

Figure 26. Clusters derived from different zinnwaldite–quartz mixtures. *** most representative sample of cluster, + are the most different ones in the clusters.

The XRD patterns used for PLSR calculation and the different XRDs in the clusters are given in Figure 27.

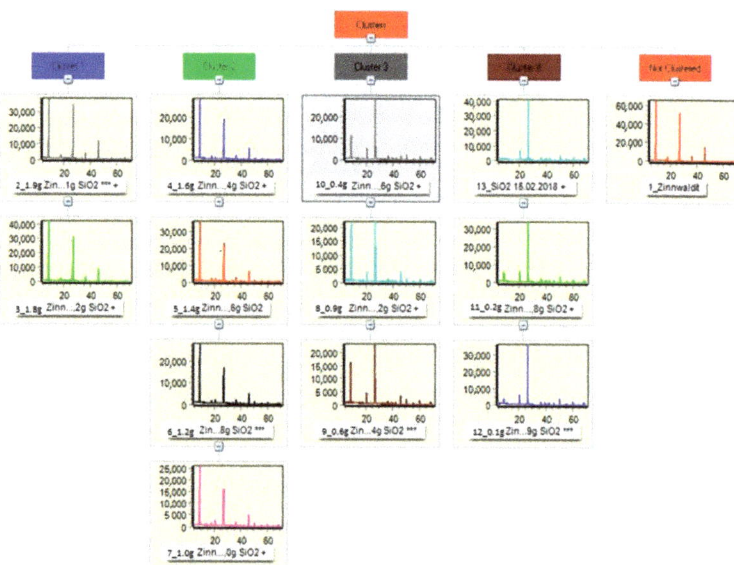

Figure 27. XRD patterns of the different cluster in zinnwaldite–quartz mixtures.

3.7. Analysis of Complex Lithium Ores and Process Mixtures

Different compositions of complex ores from different parts of the world were also analyzed and compared (Figure 28). These mixtures were all put into separate clusters by cluster analysis due to their mineralogical variances.

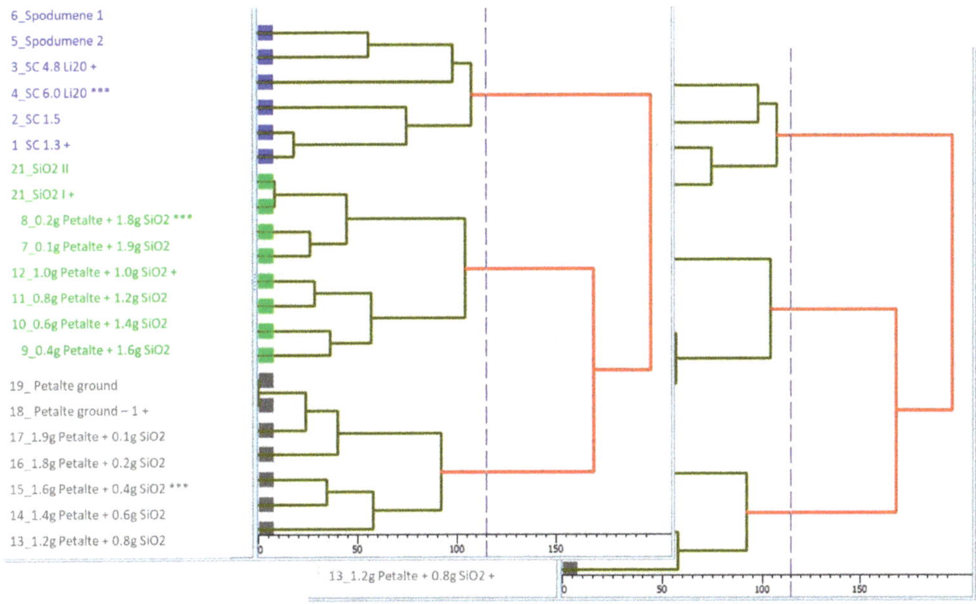

Figure 28. Dendrogram showing lithium ores from different parts of the world arranged in different clusters.

Some more ores from different parts of the world were clustered into lithium–rich mineral ores and those having higher contents of quartz. The different main lithium minerals which composed these ores could also be clustered. The different lithium ore minerals were arranged in different clusters due to their varying XRD-patterns, representing also different Li_2O-contents. A separation of the different lithium ores to different mines is possible. As the mica minerals zinnwaldite and lepidolite show close related and similar XRD patterns and therefore their main peaks do fall in the same cluster areas (Figure 29). In geological surroundings of these lithium ores normally only one type of mica is present and can then be shown, included and clustered separately here.

The following five lithium containing clusters could be separated Cluster 1: MICA: lepidolite–zinnwaldite, Cluster 2: triphylite, Cluster 3: petalite, Cluster 4: amblygonite, and Cluster 5: spodumene (Figure 29).

3.8. Investigation of Complex Ores Composed of Spodumene, Mica (Lepidolite), Quartz and Feldspar

Different binary, ternary, and quaternary mixtures in relevant compositional mineralogical variances from different occurrences were investigated. In Figure 29, the different mixtures are shown to fall into separate clusters from mixtures containing different minerals. The clusters represent XRD patterns with high contents of the lithium minerals, but also some with ternary mixtures including gangue minerals (spodumene, quartz and mica). The combinations of binary mixtures of quartz-spodumene, quartz-feldspar and feldspar-spodumene out of the quaternary tetrahedron including mica is given in Figure 30, represented by their PLSR calibration curves.

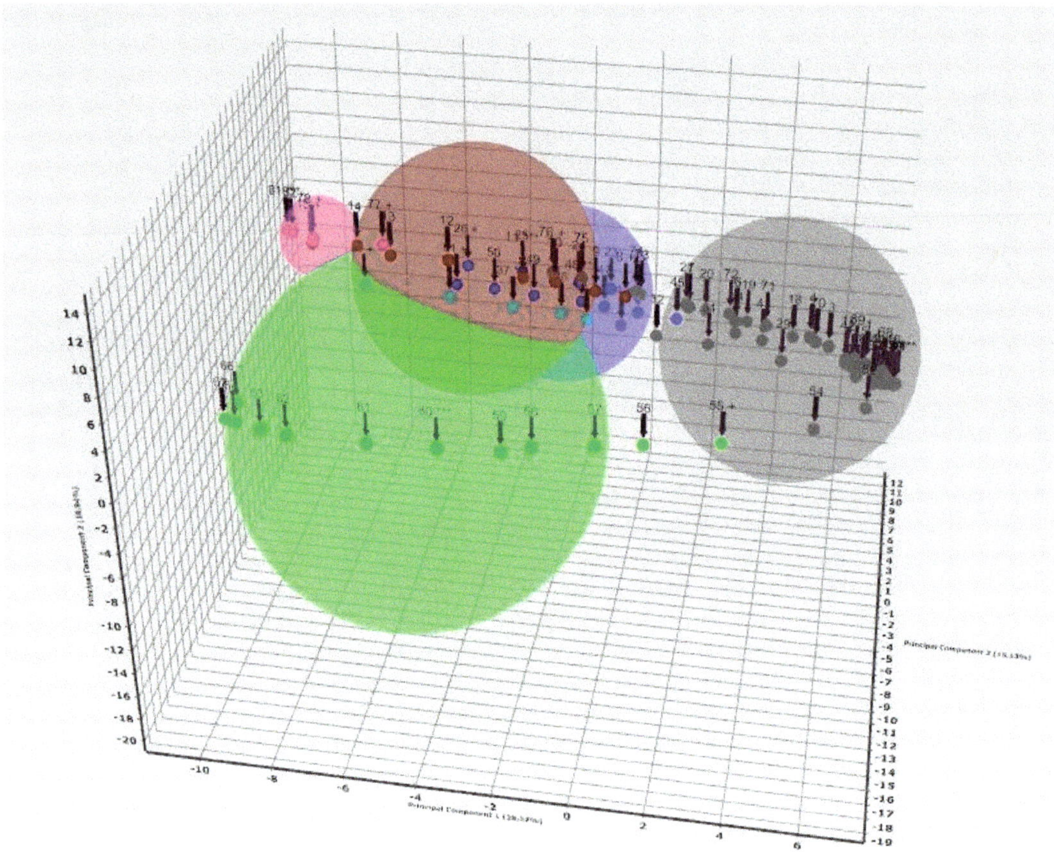

Figure 29. Five Clusters of different lithium mineral ores from different parts of the world with the main minerals separated into different clusters. *** most representative XRD, + most different XRD's in the cluster.

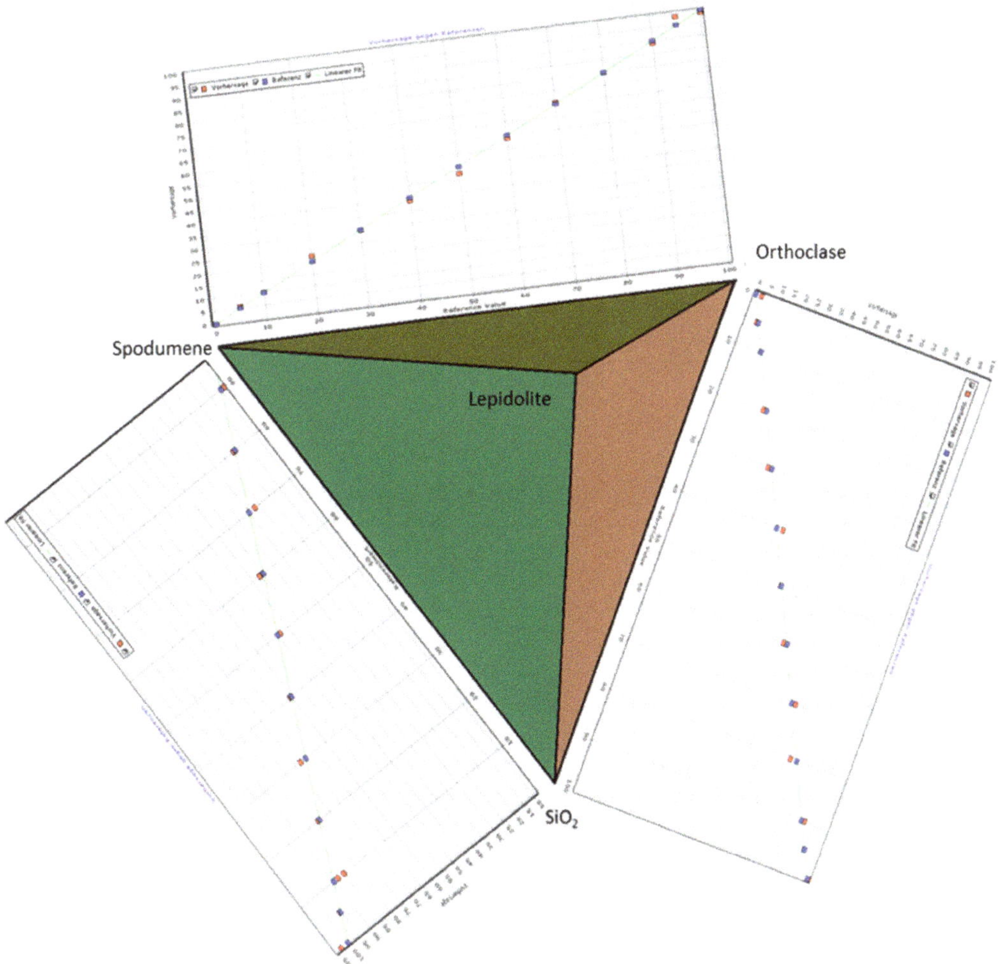

Figure 30. Calibration curves of the three binary mixtures in the system quartz-feldspar-spodumene arranged along the triangle spodumene-quartz-orthoclase.

It is also possible to treat complex mixtures by XRD, using statistical methods. The three calibration curves for binary mixtures obtained by PLSR techniques are given in Figure 30. The binary and ternary calibration curves including orthoclase, spodumene and quartz are given in Figure 31a–c. For ternary mixtures the calibration curves can be derived also by PLSR (Figure 31d) and are based on increasing spodumene contents. In Figure 32 the derived 4 clusters of the different mineral mixtures (binary and ternary) are given.

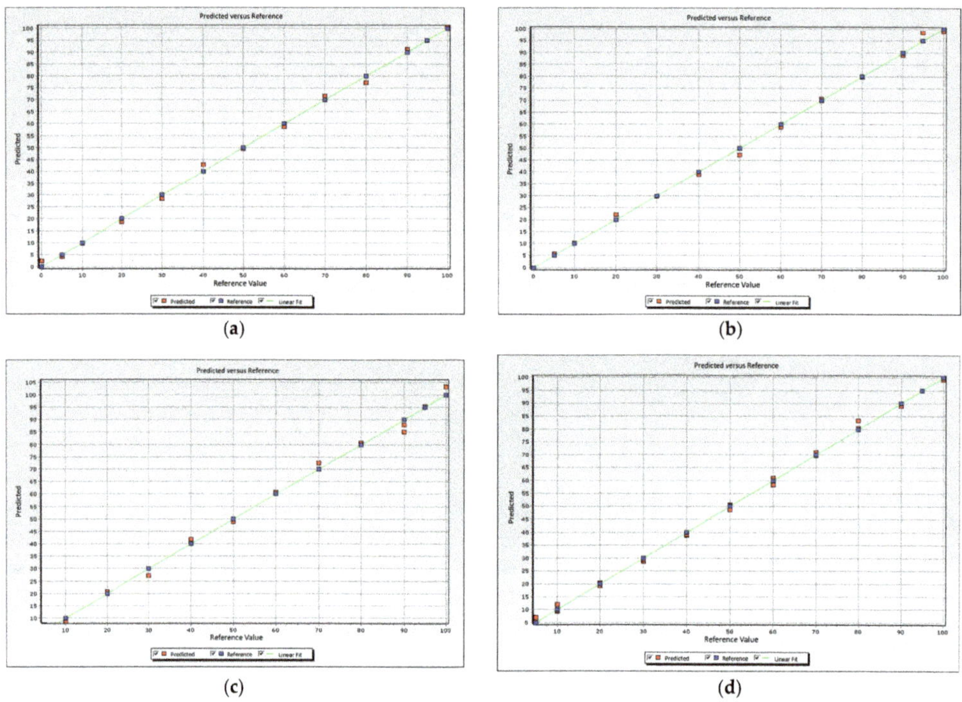

Figure 31. (a) Calibration curve of binary system orthoclase–quartz, (b) calibration curve of binary system orthoclase–spodumene, (c) calibration curve of binary system spodumene-quartz, and (d) calibration curve for spodumene in ternary mixtures orthoclase–spodumene–quartz.

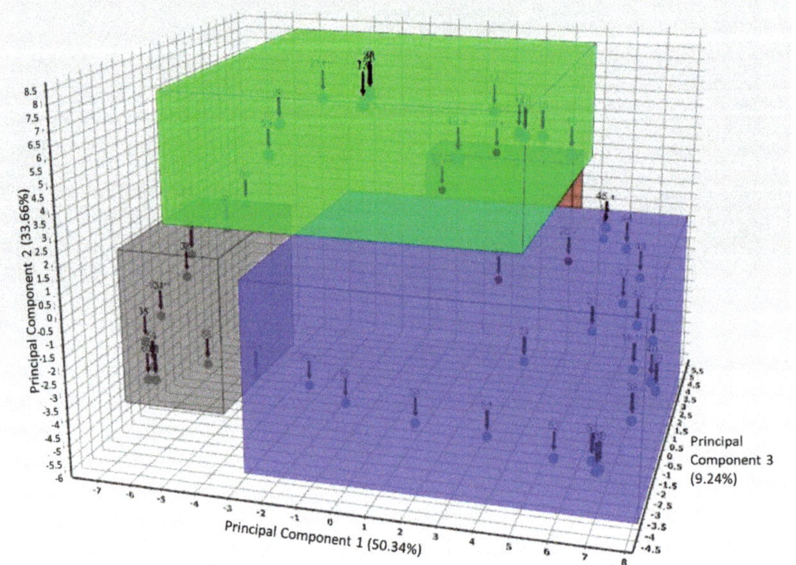

Figure 32. Separate clusters of three binary and one ternary mixture of minerals orthoclase, quartz, and spodumene.

In Figure 32 the four clusters of spodumene-rich, quartz-rich, orthoclase-rich, and ternary mixtures can be calculated.

When this system is opened up by the addition of all of the other already described binary mixtures containing lithium minerals, the clusters containing different, but definite mineral associations can be derived as given in Figure 32. These arrangements are basically due to the relevant lithium-containing mineral from which the Li_2O content can be calculated.

By the aid of this technique the different mineral associations can be separated and different mixtures of lithium containing minerals are separated (Figure 33). Only lepidolite and zinnwaldite, due to their similar XRD patterns (typical mica XRD patterns), are difficult to distinguish from each other. In practical measurements from different occurrences worldwide, not all these different combinations will be available, which concludes that these calculations for lithium minerals are even more simple to obtain.

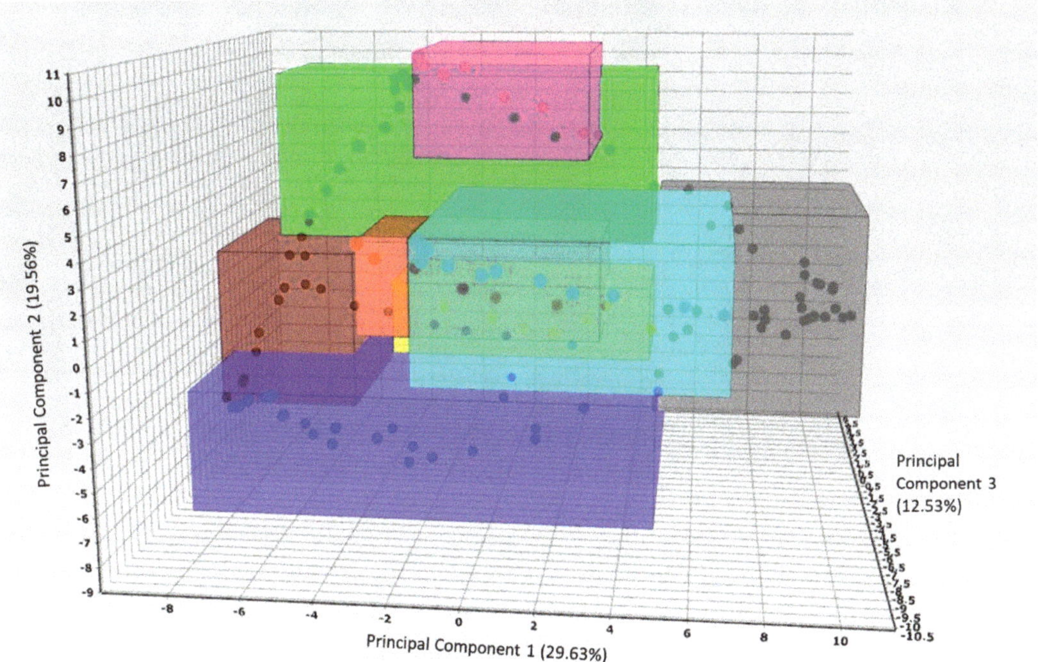

Figure 33. PCA analysis of binary, ternary, and quaternary mixtures containing all six different lithium minerals, and orthoclase and quartz, one cluster is formed containing two mica minerals.

A definite content of lithium oxide can be derived from all these mixtures, making a wet chemical analysis of lithium redundant.

4. Investigation of Special Hard Rock Ores and Brines—Industrial Case Studies on Complex Ores

4.1. Hard Rock Lithium Deposits—Characterization of Raw and Processed Materials

Lithium hard-rock ores are extracted either using open-pit or underground mining. The economically valuable fraction of lithium hard rock deposits is represented by spodumene, apatite, lepidolite, tourmaline, and amblygonite, of which spodumene is the most common lithium-bearing mineral. The gangue fraction of lithium hard-rock deposits typically consists of quartz, feldspar and other silicates.

Figure 34 describes the basic steps to extract lithium carbonate from pegmatite hard rock deposits. After crushing and separation of non-valuable ore, α-spodumene is concentrated by grinding and flotation. Byproducts such as quartz-feldspar sand can be used as filler material in various applications. α-Spodumene is calcined into β-spodumene to enable further processing towards lithium carbonate or hydroxide. Leaching, dewatering, removal of impurities, crystallization, and filtration transform β-spodumene into lithium carbonate or hydroxide. The byproduct analcime, a porous zeolite, can be used in several applications during manufacturing of ceramics, cement, and asphalt.

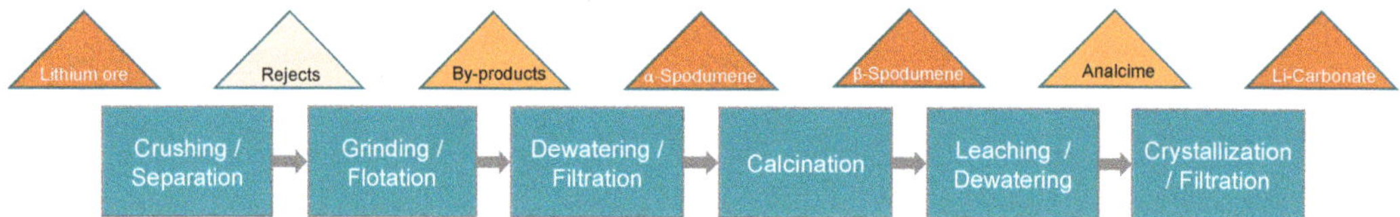

Figure 34. Schematic flow sheet for production of lithium carbonate from hard rock deposits.

The different minerals of hard rock lithium ores have different properties during processing and influence the efficiency of flotation and calcination. Therefore, frequent mineralogical monitoring allows to sort and blend different ore grades to ensure a consistent quality for processing, minimize costs for reagents and reach optimal recovery rates.

Fourteen samples including raw ores as well as processed material (flotation and calcination) were mineralogically analyzed for this case study. A benchtop X-ray diffractometer Aeris Minerals was used for the case study. Five minutes of measurement time per samples was chosen to enable fast and frequent monitoring in a process environment [15].

Prior to phase identification and quantification, cluster analysis of the raw ore samples ($n = 9$) was trialed to investigate the potential of fast ore sorting based on mineralogical in formation from XRD measurements. Two cluster (groups of similar samples) could be identified as well as two outliers (not belonging to any cluster), (Figure 35). Later mineral quantification will show, that these two clusters reflect high- and low-grade ores and can potentially be used to identify different ore domains in the mine.

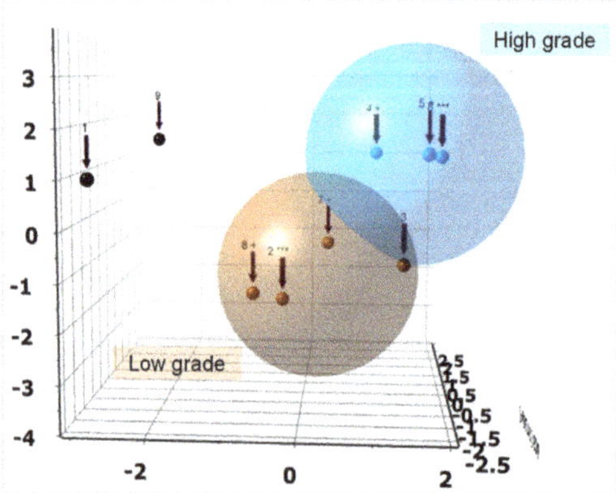

Figure 35. PCA (principal component analysis) plot with two different cluster of lithium ores representing different ore grades ($n = 9$), *** indicates the most representative scan of a cluster, + indicates the two most different scans within one cluster.

Figure 36 shows the XRD patterns of the representative raw ore sample, α- and β-spodumene concentrates and corresponding tailings.

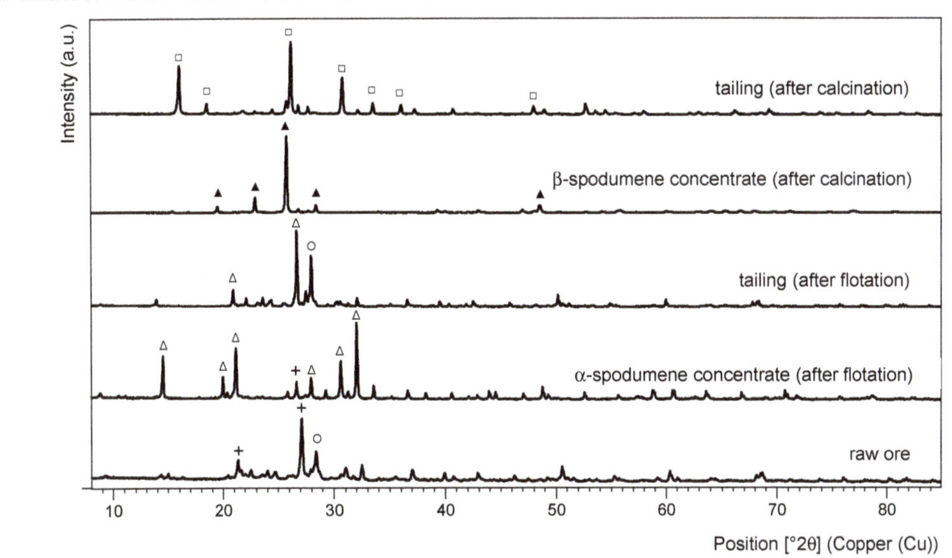

Figure 36. XRD patterns from lithium ore, concentrates and tailing after flotation and calcination, Δ—α-spodumene, ▲—β-spodumene, O—albite, +—quartz, □—analcime, modified after Norberg et al. (2020) [37].

Phase identification of the ore samples revealed a complex mineralogy. Main minerals identified are spodumene $LiAl(SiO_3)_2$, quartz SiO_2, albite $NaAlSi_3O_8$, anorthite $CaAl_2Si_2O_8$, minor amounts of lepidolite $K(Li,Al)_3(Al,Si,Rb)_4O_{10}(F,OH)_2$, orthoclase $KAlSi_3O_8$ and traces of tourmaline (elbaite) $Na(Li_{1.5}Al_{1.5})Al_6Si_6O_{18}(BO_3)_3(OH)_4$, and beryl $Be_3Al_2(SiO_4)_6$. The major peaks of the main phases are marked in the Figure 36.

The Rietveld method was applied to quantify the mineral concentration of the lithium ore samples. Figure 37 shows an example of the resulting full-pattern Rietveld refinement of one ore sample and relative quantities of all available ore samples. The information about the different ratios of spodumene and gangue minerals can be used to define grade blocks in the mine and to sort and blend ores from different mineralogical domains. Further, ore mineralogy directly influences the efficiency of the flotation step and subsequently spodumene recovery rates. The types and quantities of reagents used during flotation require carefully adjustment based on the mineralogy. In addition, the presence and amount of hard minerals such as quartz allows to react on changing ore composition and enables fast feedback times to ensure optimal lifetime of processing equipment like grinding mills.

Besides the definition of ore grades, XRD can also be used to directly monitor the efficiency during downstream processing. Figure 38 shows the quantitative phase composition of α-spodumene concentrate, tailings, or rejects after flotation [44], the calcined product, and the by-product analcime. For this example, the flotation product contains 89.6% α-spodumene with the minor amounts of quartz, albite, anorthite, and traces of lepidolite, beryl, orthoclase and mainly analcime in the residue. The main fraction of the gangue minerals is separated in the rejects, which primarily consist of albite, quartz, and anorthite.

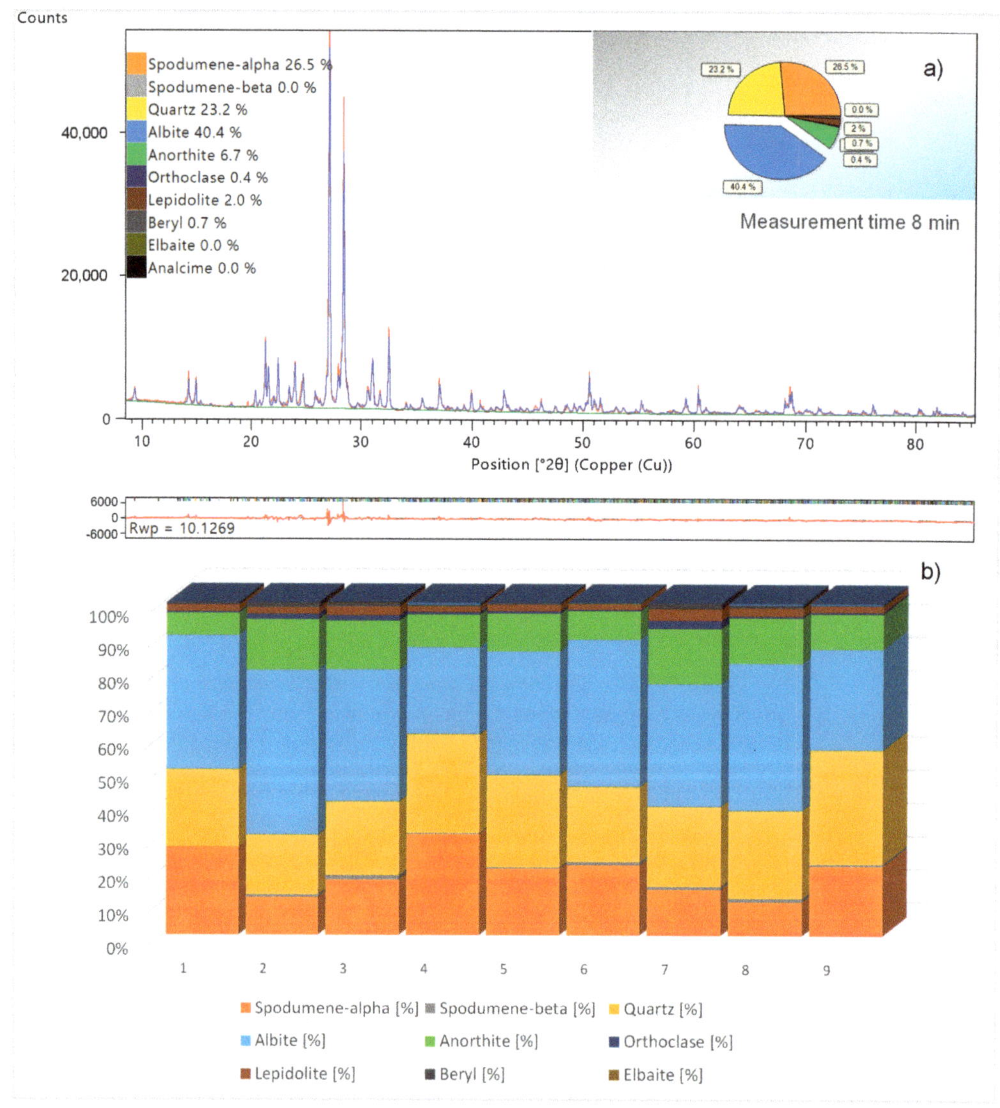

Figure 37. Example of a full-pattern Rietveld quantification of a complex hard-rock lithium ore (**a**) and variation of mineral quantities in the different products (**b**).

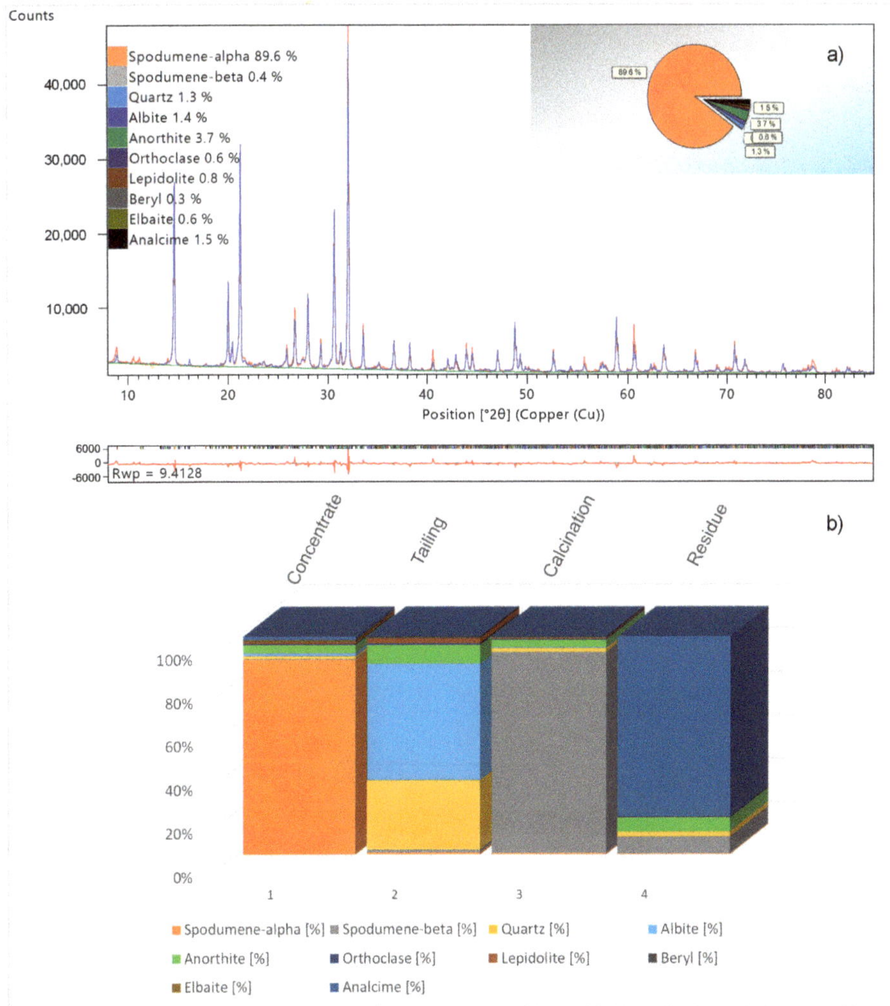

Figure 38. Example of a full-pattern Rietveld quantification of a spodumene concentrate (**a**) and composition of concentrate, tailing, calcination product and residue (**b**).

The efficiency of the transformation from α-spodumene to β-spodumene can be monitored by XRD. The two different modifications consist of different crystal structures and show different XRD patterns. In the example in Figure 38, over 92% of the calcined concentrate is transformed into β-spodumene. Impurities are minor amounts of anorthite and quartz. The corresponding by-product analcime still has a residue of β-spodumene that might require adjustments during the calcination process to further increase recovery rates. To summarize the use of XRD for monitoring hard rock lithium ores and processing materials enables easy and fast definition of mineralogical domains in the mine. Cluster analysis can be used as a tool to distinguish fast and easy between different ore grades and allows the definition of grade blocks (Figure 39) and the sorting and blending of ores based on the mineralogical composition. XRD also enables process control during the extraction of β-spodumene from the ore feed.

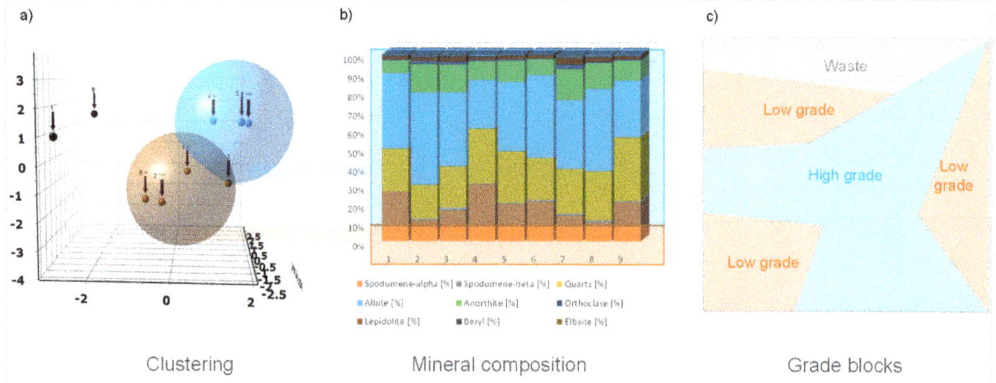

Figure 39. XRD cluster analysis (**a**), corresponding mineral quantities (**b**) and visualization of a theoretical grade block model (**c**) and mineralogical domains of a lithium hard rock deposit.

4.2. Lithium Brines—Mineralogical Characterization of Salts

Besides hard rock lithium deposits, lithium brines are also a main primary resource of lithium with 60% of the global identified reserves [46]. Salars, dried salt lakes, hold 78% of the lithium brine reserves. The process to produce lithium from lithium brines revolves around concentrating the brines up to 6% lithium and removing the impurities subsequently (Figure 40). Careful monitoring of the different salts crystallizing out of a brine helps to adjust extraction of the lithium.

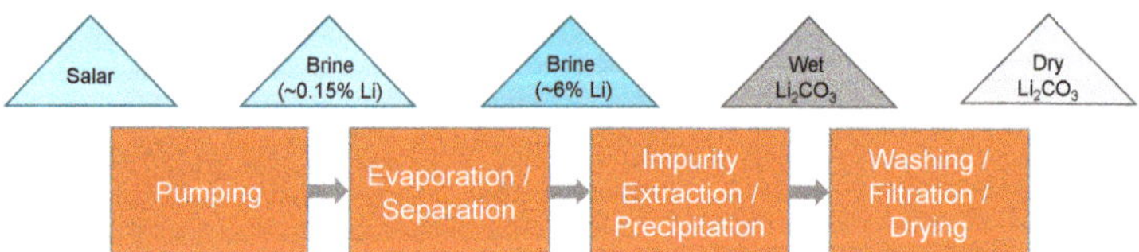

Figure 40. Simplified flow sheet for production of lithium carbonate from brines.

Forty-three samples from a lithium brine were used to cluster, identify and quantify the different mineral phases. Focus of this study was to find a fast way to characterize different salt domains. Due to the hygroscopic behavior of some salt phases, special attention was also put on the optimal sample preparation for the XRD measurements. Lithium salts like Li-Carnallite or Li-Sulfates are hygroscopic under ambient conditions and changing from solid to liquid within minutes. Although XRD offers short measurement times of about 5 min, the hygroscopic behaviour challenges phase identification and quantification. Figure 41 shows the measurement of one sample form a brine containing lithium salts. Extremely short measurement times of 1 min were chosen to follow to decrystallization of Li-Carnallite. Within 10 min all Li-Carnallite peaks disappear completely whereas the other salts are still stable. To avoid decrystallization a capton foil was used on top of the sample during the measurement (Figure 42). After 10 min, Li-Carnallite still appears crystalline and all corresponding peaks have the same intensities.

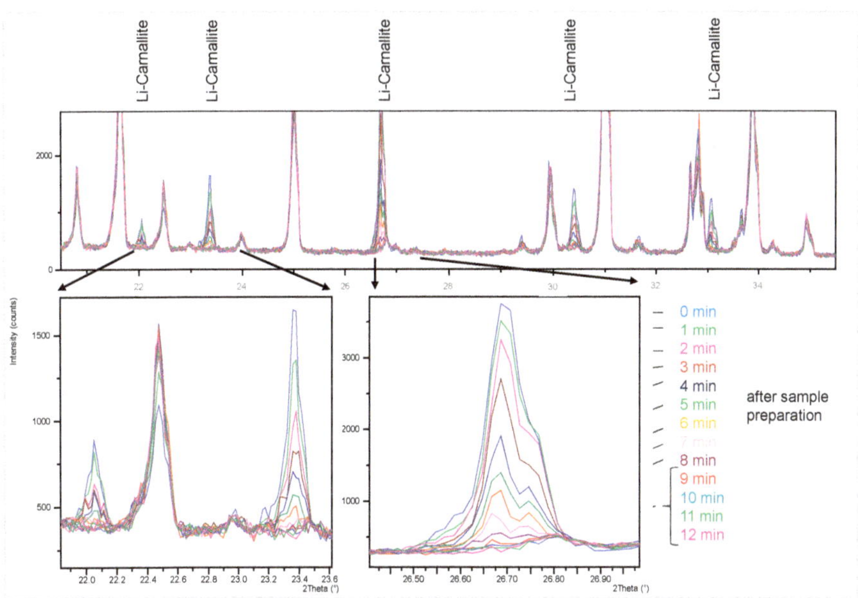

Figure 41. XRD measurement of one brine sample containing lithium carnallite using an open sample holder under ambient conditions, sample measured in 1 min intervals directly after preparation.

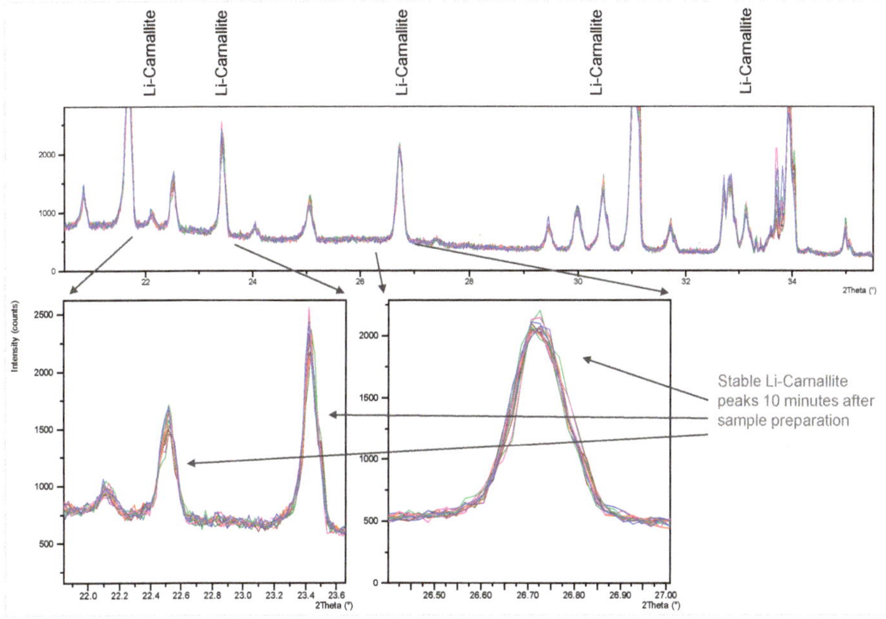

Figure 42. XRD measurement of one brine sample containing lithium carnallite using capton foil on top of the sample, measured in 1 min intervals directly after preparation.

To investigate a fast method to identify and monitor different types of salts with different mineralogical composition, cluster analysis was applied using all 43 available samples. Figure 43 shows a principal component analysis (PCA) with corresponding dendrogram, identifying seven distinct clusters. During subsequent quantification of all samples the different cluster could be connected to a certain mineral composition. Not only the chlorides, sulfates, and lithium containing salts could be separated, but also samples with different ratios of the minerals (e.g., halite and sylvine).

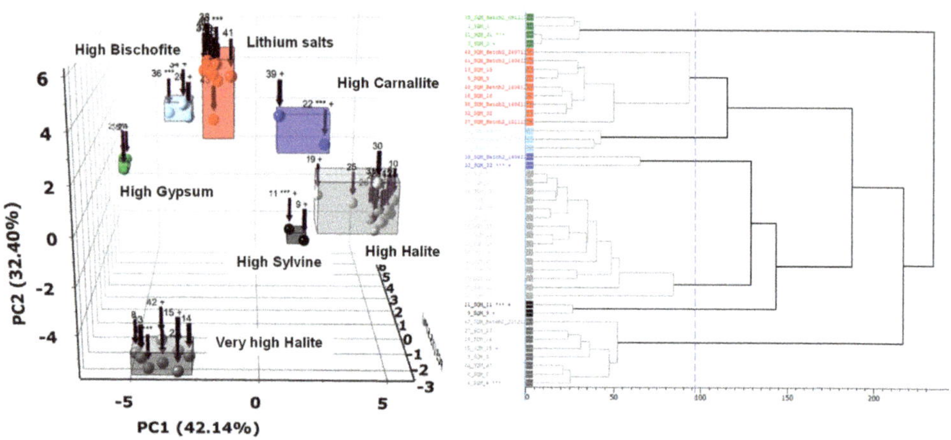

Figure 43. Cluster analysis identifying different clusters of evaporites from a lithium brine deposit (*n*-43) and the relevant dendrogram, *** indicates the most representative scan of a cluster, + indicates the two most different scans within one cluster.

For the set of 43 samples, 13 different mineral phases are identified using XRD, Table 4. Main lithium containing phases are Li-Carnallite and Li-Sulfate Monohydrate.

Table 4. Minerals identifies in the brine samples investigated.

Mineral	Formula
Halite	$NaCl$
Sylvine	KCl
Carnallite	$KMgCl_3 \cdot 6H_2O$
Bischofite	$MgCl_2 \cdot 6H_2O$
Chloromagnesite	$MgCl_2$
Anhydrite	$CaSO_4$
Gypsum	$CaSO_4 \cdot 2H_2O$
Syngenite	$K_2Ca(SO_4)_2 \cdot H_2O$
Polyhalite	$K_2MgCa_2(SO_4)_4 \cdot 2H_2O$
Kainite	$KMg(Cl,SO_4) \cdot 2.75H_2O$
Picromerite (Schoenite)	$K_2Mg(SO_4)_2 \cdot 6H_2O$
Li-Sulfate Monohydrate	$Li_2SO_4 \cdot H_2O$
Li-Carnallite	$LiMgCl_3 \cdot 7H_2O$

Quantification of the different mineral phases in all samples was applied using the Rietveld method. Figure 44 shows an example of a Rietveld refinement quantifying eight crystalline mineral phases.

The case study demonstrates that XRD can be used to monitor different salts of lithium brines. Special sample preparation is required due to the hygroscopic of the lithium salts. Cluster analysis is a fast method to identify different mineralogical compositions of the salts. Due to measurement times of about 5 min and quantification of the total mineral content (crystalline phases) it is possible to use XRD as process control method to monitor the mining and processing of lithium brines.

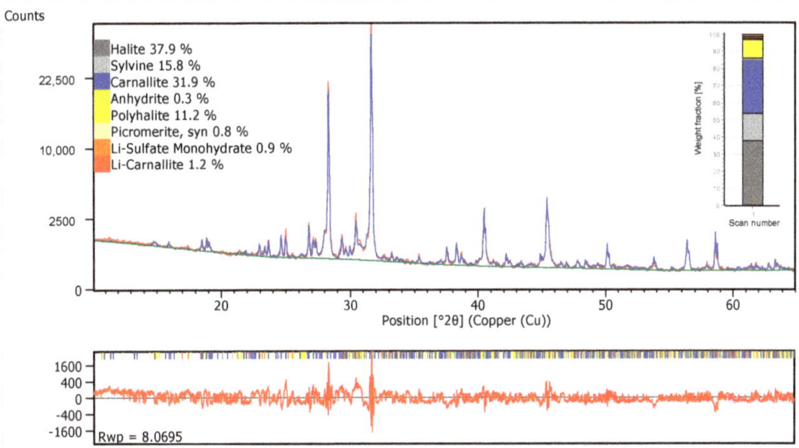

Figure 44. Example of a full-pattern Rietveld quantification of a lithium ore from a brine.

5. Conclusions

By using the XRD method for the quantitative determination of minerals, followed by a recalculation of the Li_2O-content, it is possible to replace the wet chemical analysis and obtain contents of lithium increasingly fast. It could be proved that Li_2O-contents down to 0.1% can be reliably determined in the presented mixtures after definition of the occurring lithium mineral in the sample. The method must be adapted to more complex mixtures. Nowadays besides the determination of lithium in primary ores the determination of lithium in lithium mineral leaching residues can also be of interest. It can be summarized as follows:

1. Quantitative lithium determination by mineral quantification is possible;
2. Definite mineral composition must be known for calculations of contents;
3. Rapid and fast lithium determination in mineral mixtures by XRD is possible;
4. XRD quantification method is easily applied for simple mineralogical compositions of minerals and can replace wet chemical analysis;
5. XRD methods for typical ternary mineral systems can already be successfully applied;
6. XRD methods and interpretations combined with statistical treatment methods can be applied in practical applications:
    (A) Rietveld quantification—no calibration curve necessary
    (B) Partial Least Squares—refinement with calibration curves—rapid and reliable for lithium concentrates
    (C) Clustering of different lithium mineral compositions—identifies different lithium ore qualities
    (D) Classic lithium content determination using chemical methods—time consuming but useful for referencing
7. More complex ores and brines can be treated by XRD and useful results are obtained methodology is more complicated;
8. Determination of lithium in brines is more complicated due to lower lithium contents and due to complex mineralogy;
9. For detailed mineralogical determinations a more sophisticated Rietveld analysis is useful for multi-mineral mixtures and more detailed results.

6. Summary

Thus, also XRD technique can be used for the determination of lithium in lithium ores as fast, quantitative and reliable method. The detection limits for the different minerals are summarized in Table 5. In concentrates however, mainly the higher lithium concentrations are of interest for further processing.

Table 5. Detection of lithium in six different lithium-minerals and their detection limits.

Mineral Name	Detection Limit (% of Li_2O)	Detection Limit (Mineral Content in %)	Content Li_2O in % (Ideal Composition)
Triphylite	0.1%	<1	9.47
Spodumene	0.1%	1–2	8.03
Amblygonite	0.1%	1	7.4
Lepidolite	<0.1%	<1	7.7
Zinnwaldite	0.1%	1	3.42
Petalite	0.1%	1	4.5

The procedures followed in this type of analysis, using primarily the quantitative mineral content and performing a recalculation of the lithium content in these minerals, can be used on the different raw materials. A short description is given schematically in Figure 45.

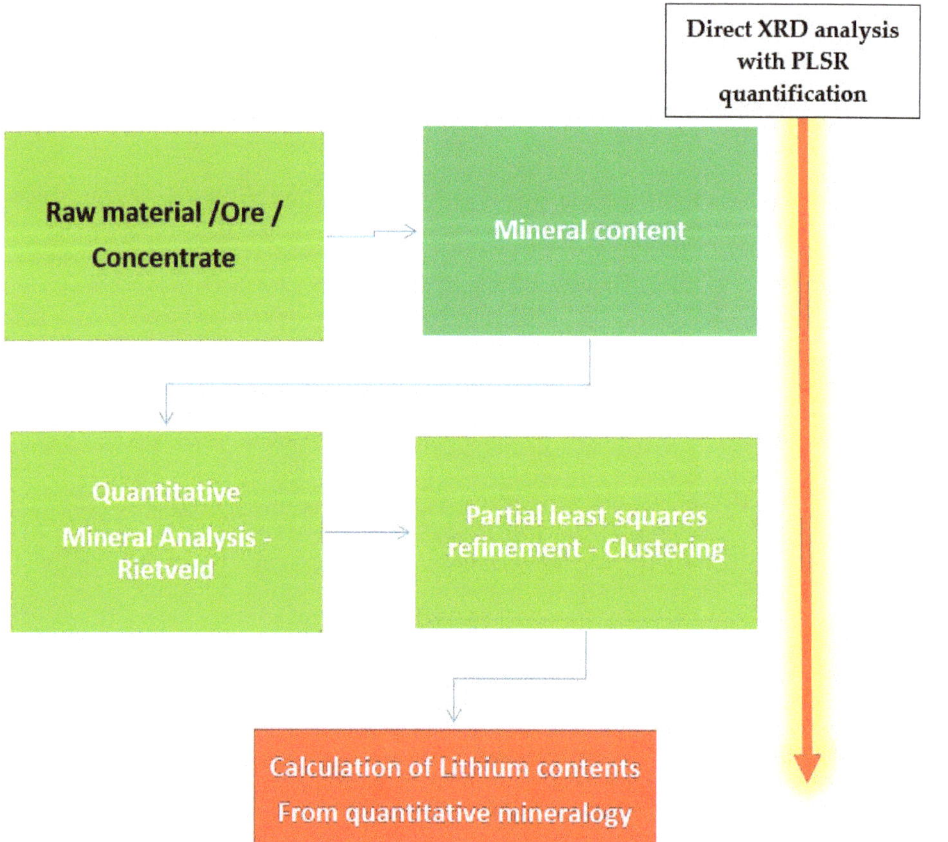

Figure 45. Schematic procedure of lithium determination in lithium ores by XRD.

Some examples of typical concentrates of lithium minerals with quartz were analyzed and their determination results based on the PLSR curves are given in Table 6 and Figure 46. The measured results of different mixtures can be determined rather exactly, sometimes some difference can be observed (maximum 2% of mineral contents). But mainly, very precise contents could be obtained.

Table 6. Examples of PLSR results of lithium ores based on definite initial values given and their measured results from XRD.

Initial Value (%)	Measured Value (%)					
	Spodumene	Triphylite	Petalite	Montebrasite	Zinnwaldite	Lepidolite
10	9.88	8.02	10.34	10.13	10	9.92
50	49.89	50.14	51.32	49.99	49.85	49.93
80	80.02	79.97	80.33	80.02	80.26	79.63
95	95.22	94.75	94.85	95.02	94.93	94.55
97.5			96.1			96.89
99	97.01	97.03				
100					99.7	

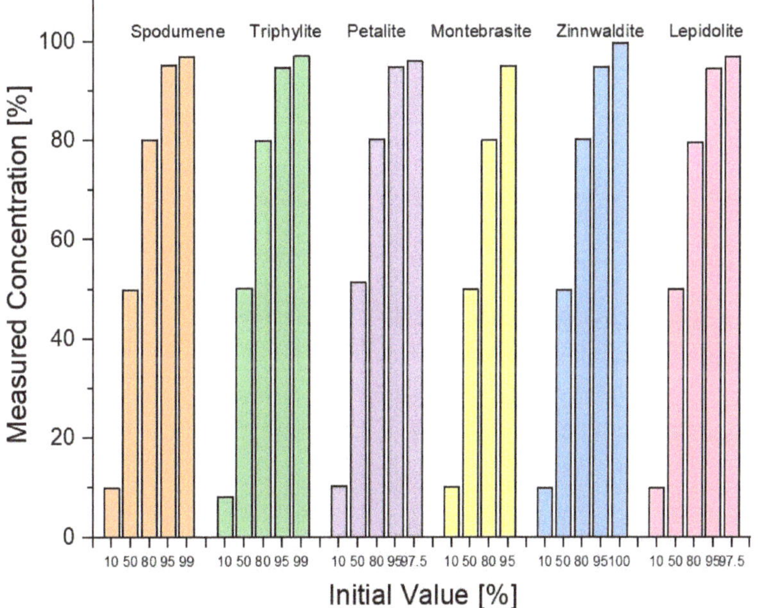

Figure 46. Measurements and results of typical concentrates of the lithium minerals and the measured contents determined by XRD.

Therefore, X-ray powder diffraction (XRD) is an established, fast, and accurate mineralogical method providing valuable information for mining and beneficiation of lithium hard-rock deposits. Accurate mineralogical analysis during mining and processing of lithium ores allows efficient mine planning, ore sorting and blending as well as optimizing the different steps of beneficiation and extraction. XRD is beneficial to optimize the use of expensive reagents in the flotation cells. Mineralogical analysis of concentrates and tailings during flotation and calcination of lithium hard-rock treatment allows fast counteractions on changing ore grades or process conditions and subsequently increases the recovery rate of lithium minerals. Modern diffraction instruments are fast, accurate and compact in size. Their infrastructure can be easily implemented in the process flow at the beneficiation plant, including full automation, from sampling to results reporting into a centralized plant control system. Compact XRD instruments can even be a part of a mobile container lab, located directly at a mine site. Automated sample preparation in combination with intuitive measurement flow setup and fully automated analysis significantly low is the entry level, required to operate a modern XRD instrument in the efficient manner.

Author Contributions: Conceptualization first part, H.P., second part, U.K.; methodology, H.P. and U.K.; writing and editing done by both authors; investigation done at University of Halle and Panalytical analytical laboratories in Almelo/Netherlands. All authors have read and agreed to the published version of the manuscript.

Funding: This research was not funded externally but resources of University of Halle/mineralogy were used.

Data Availability Statement: All data were gathered and treated at mineralogy laboratories at University of Halle and Panalytical laboratories in Almelo.

Acknowledgments: The financial help by providing analytical instruments at university of Halle and the continuous effort of Frau Kummer in the mineralogical laboratories is gratefully acknowledged.

Conflicts of Interest: The authors declare no conflict of interest.

References

1. Chaussidon, M.; Robert, F. Lithium nucleosynthesis in the Sun inferred from solar wind 7Li/6Li ratio. *Nature* **1999**, *402*, 270–273. [CrossRef]
2. Liao, B.; Chen, J. The application of cluster analysis in X-ray diffraction phase analysis. *J. Appl. Cryst.* **1992**, *25*, 336–339. [CrossRef]
3. Penniston-Dorland, S.; Liu, X.-M.; Rudnick, R.L. Lithium isotope geochemistry. *Rev. Mineral. Geochem.* **2017**, *82*, 165–217. [CrossRef]
4. Pöllmann, H.; König, U. Determination of Lithium Concentrations in Different Ores Determined by X-ray Diffraction and Statistical. In Proceedings of the International Mineralogical Association IMA 2018 Meeting, Melbourne, Australia, 13–17 August 2018.
5. Tomascak, P.B.; Magna, T.; Dohmen, R. *Advances in Lithium Isotope Geochemistry*; Springer International Publishing: Cham, Switzerland, 2016; p. 195.
6. Bibienne, T.; Magnan, J.-F.; Rupp, A.; Laroche, N. From mine to mind and mobiles: Society's increasing dependence on lithium. *Elements* **2020**, *16*, 265–270. [CrossRef]
7. Evans, K. Lithium. In *Critical Metals Handbook*; Gunn, G., Ed.; American Geophysicals Union: Washington, DC, USA, 2014; pp. 230–260.
8. Tadesse, B.; Makuei, F.; Albijanic, B.; Dyer, L. The beneficiation of lihtium minerals from hard rock ores: A review. *Miner. Eng.* **2019**, *131*, 170–184. [CrossRef]
9. Benson, T.R.; Coble, M.; Rytuba, J.J.; Mahood, G.A. Lithium enrichment in intracontinental rhyolite magmas leads to Li deposits in caldera basins. *Nat. Commun.* **2017**, *8*, 270. [CrossRef] [PubMed]
10. Bowell, R.J.; Pogge von Strandmann, P.A.E.; Grew, E.S. Lithium: Less is more. *Elements* **2020**, *16*, 4.
11. Bowell, R.; Lagos, L.; de los Hoyes, C.R.; Declercq, J. Classification and characteristics of natural lithium resources. *Elements* **2020**, *16*, 259–264. [CrossRef]
12. Garrett, D.E. Their deposits, processing. Use and properties. In *Handbook of Lithium and Natural Calcium Chloride*; Elsevier: Amsterdam, The Netherlands, 2004; p. 476.
13. Kesler, S.E.; Gruber, P.W.; Medina, P.A.; Keoleian, G.A.; Everson, M.P.; Wallington, T.J. Global lithium resources: Relative importance of pegmatite, brine and other deposits. *Ore Geol.* **2012**, *106*, 55–69. [CrossRef]
14. Gourcerol, B.; Gloaguen, E.; Melleton, J.; Tuduri, J.; Galiegue, X. Re-Assessing the European lithium resource potential—A review of hard-rock resources and metallogeny. *Ore Geol. Rev.* **2019**, *109*, 494–519. [CrossRef]
15. London, D. *Pegmatites. The Canadian Mineralogist. Special Publication 10*; Mineralogical Association of Canada: Quebec City, QC, Canada, 2008.
16. London, D. Reading pegmatites: Part 3—What lithium minerals say. *Rocks Miner.* **2017**, *92*, 420–425. [CrossRef]
17. Schmidt, M. *Risikobewertung Lithium*; DERA(Deutsche Rohstoffagentur) Rohstoffinformation: Berlin, Germany, 2017.
18. Grew, E.S. The minerals of lithium. *Elements* **2020**, *16*, 235–240. [CrossRef]
19. Grew, E.S.; Hystad, G.; Toapanta, M.P.; Eleish, A.; Ostroverkhova, A.; Golden, J.; Hazen, R.M. Lithium mineral evolution and ecology: Comparison with boron and beryllium. *Eur. J. Mineral.* **2019**, *31*, 755–774. [CrossRef]
20. Černý, P.; Ercit, T.S. The classification of granitic pegmatites revisited. *Can. Mineral.* **2005**, *43*, 2005–2026. [CrossRef]
21. Černý, P.; London, D.; Novák, M. Granitic pegmatites as reflections of their sources. *Elements* **2012**, *8*, 289–294. [CrossRef]
22. Rietveld, H.M. A profile refinement method for nuclear and magnetic structures. *J. Appl. Crystallogr.* **1969**, *2*, 65–71. [CrossRef]

23. Bradley, D.C.; McCauley, A.D.; Stillings, L.M. *Mineral-Deposit Model for Lithium-Cesium-Tantalum Pegmatites*; Scientific Investigations Report; U.S. Geological Survey: Reston, VA, USA, 2017; p. 48.

24. Brand, N.; Brand, C. Detecting the Undectable: Lithium by Portable XRF—Talk S3 Presentation Big Sky Montana. In Proceedings of the DXC Conference, Big Sky, MT, USA, 31 July–4 August 2017.

25. Bale, M.; May, A. Processing of ores to produce tantalum and lithium. *Miner. Eng.* **1989**, *2*, 299–320. [CrossRef]

26. Bishimbayeva, G.; Zhumabayeva, D.; Zhandayev, N.; Nalibayeva, A.; Shestakov, K.; Levanevsky, I.; Zhanabayeva, A. Technological improvement, lithium recovery methods from primary resources. *Orient. J. Chem.* **2018**, *34*, 2762–2769. [CrossRef]

27. Braga, P.; Franca, S.; Neumann, R.; Rodriguez, M.; Rosales, G. Alkaline process for extracting lithium from Spodumene. In Proceedings of the 11th International Seminar on Process Hydrometallurgy-Hydroprocess, Santiago, Chile, 19–21 June 2019; pp. 1–9.

28. Choubey, P.K.; Kim, M.-S.; Srivastava, R.R.; Lee, J.-C.; Lee, J.-Y. Advance review on the exploitation of the prominent energy-storage element: Lithium. Part I: From mineral and brine resources. *Miner. Eng.* **2016**, *89*, 119–137. [CrossRef]

29. Dessemond, C.; Soucy, G.; Harvey, J.-P.; Ouzilleau, P. Phase transitions in the α–γ–β spodumene thermodynamic system and impact of γ-spodumene on the efficiency of lithium extraction by acid leaching. *Minerals* **2020**, *10*, 519. [CrossRef]

30. Yan, Q.; Li, X.; Yin, Z.; Wang, Z.; Guo, H.; Peng, W.; Hu, Q. A novel process for extracting lithium from lepidolite. *Hydrometallurgy* **2012**, *121–124*, 54–59. [CrossRef]

31. Shankleman, J.; Biesheuvel, T.; Ryan, J.; Merrill, D. We're Going to Need More Lithium. Bloomberg Businessweek, 7 September 2017. Available online: https://www.bloomberg.com/graphics/2017-lithium-battery-future/ (accessed on 21 July 2021).

32. Dessemond, C.; Lajoie-Leroux, F.; Soucy, G.; Laroche, N.; Magnan, J.-F. Spodumene—The lithium market, resources and processes. *Minerals* **2019**, *9*, 334. [CrossRef]

33. Ross, N.L.; Zhao, J.; Slebodnick, C.; Spencer, E.C.; Chakoumakos, B.C. Petalite under pressure: Elastic behavior and phase stability. *Am. Mineral.* **2015**, *100*, 714–721. [CrossRef]

34. Schneider, A.; Schmidt, H.; Meven, M.; Brendler, E.; Kirchner, J.; Martin, G.; Bertau, M.; Voigt, W. Lithium extraction from the mineral zinnwaldite: Part I: Effect of thermal treatment on properties and structure of zinnwaldite. *Miner. Eng.* **2017**, *111*, 55–67. [CrossRef]

35. Fehr, K.T.; Hochleitner, R.; Schmidbauer, E.; Schneider, J. Mineralogy, mössbauer spectra and electrical conductivity of triphylite. *Phys. Chem. Miner.* **2007**, *34*, 485–494. [CrossRef]

36. Norberg, N.; König, U.; Bhaskar, H.; Narygina, O. Accurate mineralogical analysis for efficient lithium ore processing. In Proceedings of the ALTA 2020 Lithium & Battery Technology, Online, 26 November 2020.

37. Rietveld, H.M. Line profiles of neutron powder diffraction peaks for structure refinement. *Acta Crystallogr.* **1967**, *22*, 151–152. [CrossRef]

38. De Jong, S. SIMPLSR: An alternative approach to partial least squares regression. *Chemom. Intell. Lab. Syst.* **1993**, *18*, 251–263. [CrossRef]

39. Lohninger, H. *Teach/Me Data Analysis*; Springer: Berlin, Germany; New York, NY, USA; Tokyo, Japan, 1999.

40. Degen, T.; Sadki, M.; Bron, E.; König, U.; Nénert, G. The highscore suite. *Powder Diffr.* **2014**, *29*, S13–S18. [CrossRef]

41. Kelley, L.A.; Gardner, S.P.; Sutcliffe, M.J. An automated approach for clustering an ensemble of NMR-derived protein structures into conformational-related subfamilies. *Protein Eng.* **1996**, *9*, 1063–1065. [CrossRef]

42. König, U.; Norberg, N. New Tools for Process Control in Aluminum Industries—PLSR on XRD Raw Data. In Proceedings of the 10th International Alumina Quality Workshop, Perth, Australia, 19–23 April 2015; pp. 305–308.

43. Wold, H. Estimation of principal components and related models by iterative least squares. In *Multivariate Analysis*; Krishnaiaah, P.R., Ed.; Academic Press: New York, NY, USA, 1966; pp. 391–420.

44. Filippov, L.; Farrokhpay, S.; Lyo, L.; Filipova, I. Spodumene flotation mechanism. *Minerals* **2019**, *9*, 372. [CrossRef]

45. Groat, L.A.; Chakoumakos, B.C.; Brouwer, D.H.; Hofmann, C.M.; Fyfe, C.A.; Morell, H.; Schultz, A.J. The amblygonite LiAlPO4F-Montebrasite LiAlPO4OH solid solution: A combined powder and single-crystal neutron diffraction and solid state Li MAS, COP MAS and REDOR NMR study. *Am. Mineral.* **2003**, *88*, 195–210. [CrossRef]

46. Grosjean, C.; Miranda, P.H.; Perrin, M.; Poggi, P. Assesment of the world lithium ressources and consequenses of their geographic distribution on the expected development of the electric vehicle industry. *Renew. Sustain. Energy Rev.* **2012**, *166*, 1735–1744. [CrossRef]

Article

Value of Rapid Mineralogical Monitoring of Copper Ores

Matteo Pernechele [1],*, Ángel López [2], Diego Davoise [2], María Maestre [2], Uwe König [1] and Nicholas Norberg [1]

[1] Malvern Panalytical B.V., 7602 EA Almelo, The Netherlands; uwe.konig@malvernpanalytical.com (U.K.); nicholas.norberg@malvernpanalytical.com (N.N.)
[2] Atalaya Mining Plc, 21660 Minas de Riotinto, Spain; angel.lopez@atalayamining.com (Á.L.); diego.davoise@atalayamining.com (D.D.); maria.maestre@atalayamining.com (M.M.)
* Correspondence: matteo.pernechele@malvernpanalytical.com

Abstract: An essential operation in the mineral processing of copper ores into concentrates is blending, as it guarantees a constant feed for the flotation cells, increases metal recovery rate and reduces tailings. In this study, copper ores from Huelva province (Spain) were investigated by quantitative XRD (X-ray diffraction) methods to optimize blending and detect penalty minerals, which can affect flotation and concentrate quality. The Rietveld method in combination with cluster analysis, PLSR and more traditional chemical analysis provide a more complete and in-depth characterization of the ore and the whole process. The mineralogical monitoring can be fully automated to enable real-time decision making.

Keywords: chalcopyrite; ore blending; copper flotation; XRD; Rietveld; PLSR

Citation: Pernechele, M.; López, Á.; Davoise, D.; Maestre, M.; König, U.; Norberg, N. Value of Rapid Mineralogical Monitoring of Copper Ores. *Minerals* **2021**, *11*, 1142. https://doi.org/10.3390/min11101142

Academic Editor: Andrea Gerson

Received: 15 September 2021
Accepted: 13 October 2021
Published: 17 October 2021

Publisher's Note: MDPI stays neutral with regard to jurisdictional claims in published maps and institutional affiliations.

1. Introduction

Copper is an essential element for today's technologies, including the pursuit of a green economy and electrification of transport. Mining of copper ores has generated 20 Mt of copper in 2020 globally, and the total reserves are estimated to be 870 Mt [1]. Approximately 250 Mt of copper ores are located at the Iberian Pyrite Belt (IPB), a massive sulfide deposit, which has been mined throughout history. The mineral beneficiation of the ores into concentrates involves comminution followed by flotation, which increases the copper content by 1–2 orders of magnitude. The selection and optimization of the separation and concentration process ultimately affect the economic recovery of the copper ore.

The ore grade is defined by both chemistry and mineralogy, but the mineralogy has a more important role in the recovery rate. When dispersed in a solution, different minerals (as well as different surfaces of the same minerals) have different interfacial energies and surface properties, such as point of zero charge, wettability and chemical affinities [2]. The flotation process exploits the difference in wettability of the different surfaces in aqueous solution to separate the valuable minerals from the gangue. The addition of different pH regulators, ions or surfactants to the solid-water dispersion can selectively change the wettability of different minerals, therefore allowing a high degree of flexibility to the separation and beneficiation process.

The knowledge and control of the mineralogy during ore blending operation, therefore, have an impact on the flotation optimization and ultimately on the copper recovery. Proper blending and optimized flotation also minimize unwanted elements which can decrease the value of the concentrates, such as arsenic (As) and antimony (Sb). The surface properties of minerals also affect agglomeration during comminution. They influence the colloidal properties of a material in suspensions and the flocculation and agglomeration phenomena during dewatering operation. For instance, the presence of swelling clays has a detrimental role in the dewatering of the tailings; therefore, the mineralogy also plays an important role in water consumptions and tailing management. If present in large quantities, clay minerals can also cause pipe blocking and, therefore, halt part of the mineral processing operations.

The main characterization technique to probe the mineralogy of a material is X-ray powder diffraction (XRPD), as it detects the atomic planes that are defined by the mineral space group and unit cell atomic content. In this work, we assess the use of the XRPD technique for the rapid quantification of minerals through the flotation process. The Rietveld refinement [3] is carried out by control files, which contain a database of tens of structures together with a pre-defined refinement strategy. The precision of the analysis was assessed by comparing the results to independent elemental analysis. The capability of clustering techniques such as PCA (principal component analysis) and predictive tools such as PLSR (partial least square regression) were also evaluated and compared with classical Rietveld results.

Correlation between mineralogy and mineral processing are presented in the discussion. The fast generation of reliable data which can be used for decision making is a key component to the digitalization of the mining industry. This research on copper ore materials and their mineral processing can easily be translated to other industrial activities, where the mineralogical knowledge can spark innovation directed toward sustainable goals [4–8].

2. Materials and Methods

2.1. Materials

In this study, we characterized different copper ore blends from the Iberian Pyrite Belt and quantified the mineralogical changes occurring during the flotation beneficiation process. The samples were provided by Atalaya (Minas de Riotinto, Spain), a European mining company focused on research, exploration and processing of copper ores. The assay results are reported in Table 1 and they were obtained using inductively coupled plasma-optical emission spectrometry (ICP Avio 500, Perkin Elmer Inc., Waltham, United States) (ICP-OES). The metal zonation of the copper ore in the region (Huelva province) is mostly composed of chalcopyrite and pyrite, while the host rock is mostly made of chlorite-group minerals, sericite and silica varieties [9]. A total of 265 samples were characterized and equally divided in three categories: the blended ores used as feed for the flotation cells, the flotation concentrates and the flotation tailings.

Table 1. Average chemical composition of the available samples. #Samples represents the number of XRD samples analyzed per each material type.

Material Type	#Samples	Cu (%)	S (%)	Zn (%)	Pb (ppm)	Fe (%)	As (ppm)	Sb (ppm)	Bi (ppm)
Blends	87	0.46	5.11	0.11	157	12.3	190	58	29
Tailings	88	0.07	4.58	0.06	115	11.98	178	17	26
Concentrates	90	20.33	35.90	3.26	2800	30.36	1260	2443	230

2.2. Method

The particle sizes of the materials were below 250 μm. Sample preparation for XRD (X-ray diffraction) included grinding with a swing mill (also known as shatterbox, vibratory disc mill or pulverizer) to reduce the particle size below 60 μm and improve mineral quantification by quantitative X-ray diffraction (QXRD). Samples were prepared the for XRD analysis in a 27 mm backloading sample holder to reduce the preferential orientation.

The XRD scans were collected using the Minerals Edition of Aeris benchtop X-ray diffractometer (Malvern Panalytical B.V., Almelo, The Netherland) (40 kV–15 mA), with a goniometer radius of 145 mm, Kα = 1.79 Å cobalt-anode X-ray tube, 0.04 Soller slits, $\frac{1}{4}$ divergence slits, 23 mm mask, low beam-knife, step size 0.02° and acquisition time of 80 s/step. The Bragg-Brentano measurement covered a range of 5–80 °2θ (atomic planes from 1.39 Å to 20.51 Å), allowing the detection of clay minerals. The use of a cobalt tube avoids iron fluorescence usually observed with Cu-radiation, therefore improving penetration depth of X-rays in the sample and improving counting statistics. The use of a linear PIXcel^{1D} Medipix3 detector (Malvern Panalytical B.V., Almelo, The Netherland) with an active length of 5.54 °2θ allows a scan acquisition time of a few minutes, and the results are immediately analyzed by automatic Rietveld routines installed in the Minerals Edition of Aeris.

2.3. Analysis

The phase identification was done with the HighScore Plus software package version 4.8 (Malvern Panalytical B.V., Almelo, The Netherland) [10] and the ICDD PDF-4 database (accessed on 3 March 2021) [11]. The phase identification was performed on specific, representative XRD scans selected by cluster analysis, which will be covered in the results section. The identified mineral phases are reported in Table 2, together with their averaged weight percentages as determined by the Rietveld refinement [12]. More than 10 other minerals were identified, but are mainly present in minor amounts, e.g., tennantite, galena, arsenopyrite, dolomite, gypsum, marcasite and chalcocite. The Rietveld routines were created with HighScore Plus and executed by software RoboRiet 4.8, a dedicated implementation of Rietveld quantification for automation projects. The results were verified by executing manual Rietveld refinements in HighScore Plus.

Table 2. Average weight percentages of major minerals as determined by Rietveld refinements. The mineral variability within the same material type is reported as the absolute standard deviation (1σ) in between brackets.

| Material | Agreements Index | | Minerals Quantification (wt%) | | | | | | | |
| | | | Cu and Zn Bearing Minerals | | | | Gangue | | | |
	Rwp	GOF	Chalcopyrite	Sphalerite	Tetrahedrite	Pyrite	Quartz	Chamosite	Muscovite	Siderite
Blends	7 (1)	3.4 (0.4)	1.1 (0.3)	0.1 (0.1)	0.0 (0.0)	5 (2)	53 (4)	30 (5)	6 (2)	0.8 (0.6)
Tailings	7 (1)	3.5 (0.5)	0.2 (0.1)	0.0 (0.0)	0.0 (0.0)	5 (2)	53 (5)	31 (6)	6 (2)	0.8 (0.3)
Concent.	3.5 (0.4)	2.2 (0.2)	55 (8)	5.1 (1.6)	0.9 (0.8)	24 (9)	4 (2)	4 (2)	0.2 (0.2)	0.9 (0.3)

3. Results

3.1. Minerals Present in the Blends, Tailings and Concentrates

The major minerals with their distinctive most intense peak positions are reported in Figure 1. The material is an ore blend used as feed for flotation. Its diffraction pattern is the representative scan as defined by cluster analysis. The main diffraction peaks correspond to quartz, pyrite and phyllosilicates (chamosite and muscovite). The main copper-bearing mineral is chalcopyrite, which has several diffraction peaks in the range of 20–70 °2θ, the most intense at 34.2 °2θ, which is better observed in the diffractograms of a representative concentrate material (Figure 2). The diffraction patterns of tailings and feeds are quite similar, while the patterns of the concentrates display lower diffraction intensity of gangue minerals and higher intensity for copper, zinc and iron sulfide minerals. A list of major minerals can be found in Table 2.

The Rietveld refinement requires XRD scans of a certain quality, especially when the mineral quantification is performed in a quality control environment with automatic RoboRiet routines. The data should be characterized by features such as a low and flat background and high diffraction intensities (most intense peak with >10,000 counts). The peak position overlap of certain minerals, e.g., sphalerite with pyrite, chalcopyrite with gypsum and tetrahedrite, supports the use of Rietveld refinement as a full pattern fitting approach, whereas classical straight-line calibration methods can cause a bias in the results.

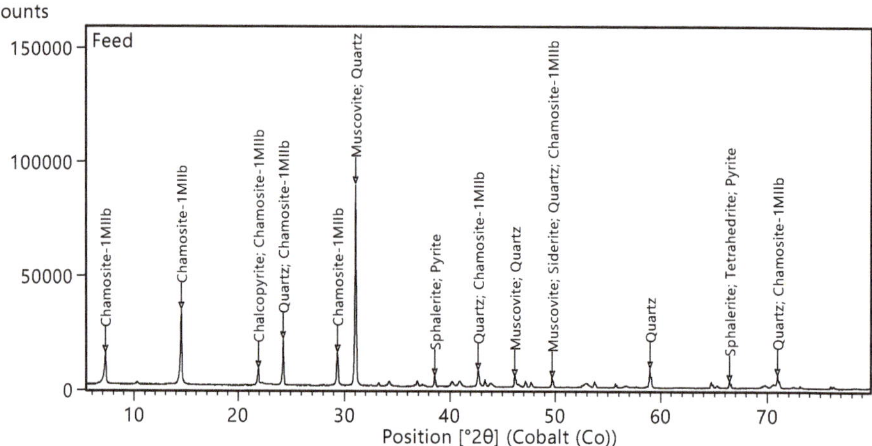

Figure 1. XRD scan of a flotation feed. The diffraction pattern of this feed is the representative scan as defined by the cluster analysis.

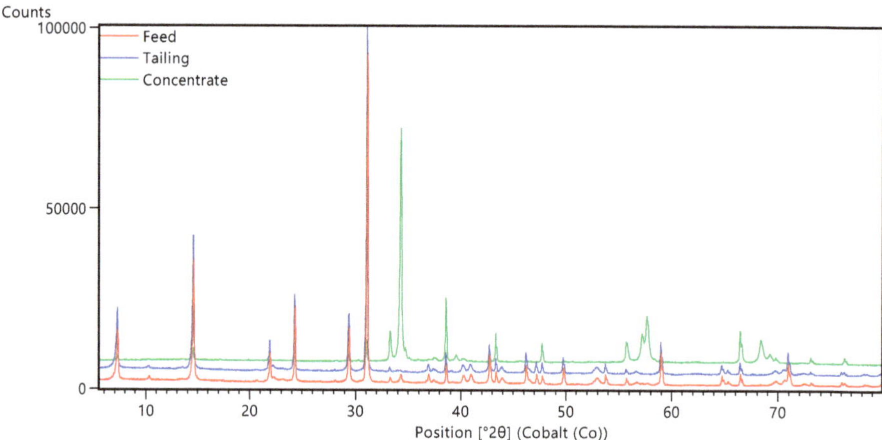

Figure 2. Characteristic XRD scans of the ore blend used as feed for the flotation cell, tailings and concentrates. XRD scans of tailings and concentrates are here vertically shifted by 3000 and 6000 counts for clarity.

3.2. Details on Chalcopyrite and Tetrahedrite Monitoring

In this section we focused on the capability of XRD techniques to detect and quantify minerals present at minor concentrations, in particular the residual chalcopyrite content in a tailing sample (Figure 3) and Bi-Sb-Pb-As sulphosalts present in a concentrate sample (Figure 4).

The visualization of the Rietveld results of different ore blends over time allows the plants to assess and control the mineral variability of the feed used in the flotation cell. The most prominent difference in the ore blends is the type of gangue mineralogy, characterized by the amount of pyrite and chamosite, Figure 5. The flotation parameters, the surfactant selection and surfactant amount can therefore be optimized based on such quantities. The variation over time of the chalcopyrite content as well as other relevant minerals included in the Rietveld analysis is shown in Figure 6. Chalcopyrite is the main copper ore mineral and it varies between 0.5 wt% and 1.9 wt%, which is well above the limit of detection, see Figure 3. The limit of detection is approximately 120 net counts, as calculated from LOD = [3 ∗ net intensity ∗ (background)$^{-0.5}$] [13]. A SNR > 10 is the minimum prerequisite to accurately quantify a mineral in a mixture. The accuracy of the Rietveld results depends also on (1) particles statistic, (2) degree of overlap with peaks from other phases, (3) preferential orientation, (4) overgrinding and (5) microabsorption.

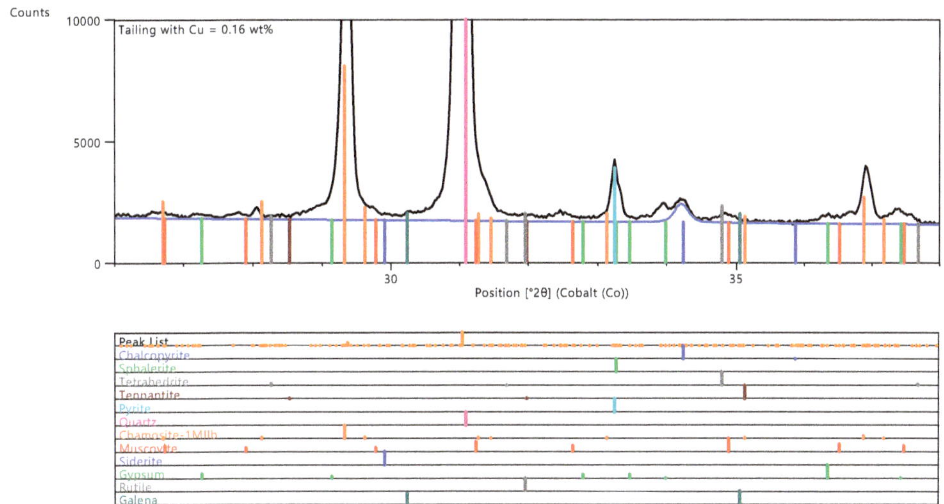

Figure 3. XRD pattern of a tailing sample with 0.16 wt% of copper in the range 26–38 °2θ. Upper curve represents measured data, lower curve in blue represents Rietveld contribution of 0.45 wt% chalcopyrite. Vertical bars are peak positions of crystalline phases.

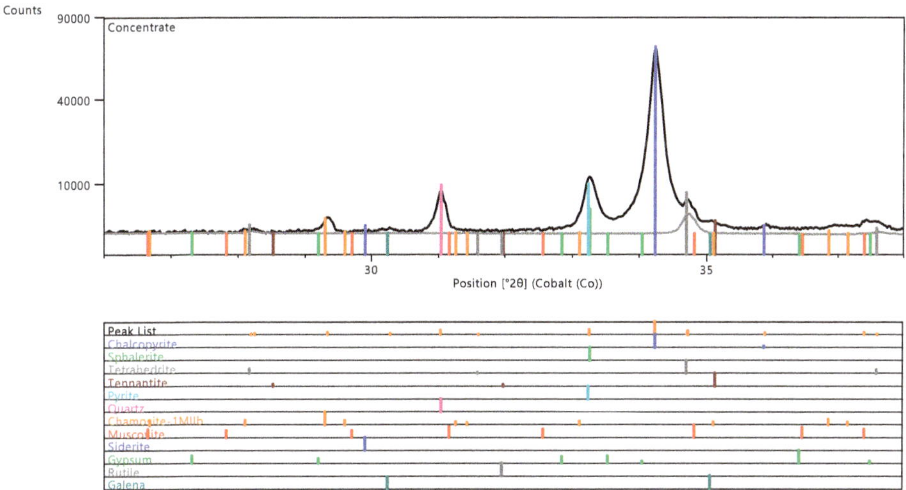

Figure 4. XRD pattern of ore concentrate in the range 26–38 °2θ. Upper curve represents measured data, lower curve in gray represents Rietveld contribution of 1.6 wt% of tetrahedrite. Vertical bars are peak positions of crystalline phases.

For the tailing sample, the copper concentration is Cu = 0.16 wt% as determined by ICP, while the chalcopyrite is 0.45 wt% as determined by automatic Rietveld refinement. Chalcopyrite contains 34.6% of copper and the back-calculated copper content from Rietveld analysis is 0.16 wt%, in agreement with ICP results. Vertical bars show diffraction peaks of chalcopyrite $CuFeS_2$, including its most intense peak at 2θ = 34.2°. The latter peak is due to the diffraction of Kα = 1.79 Å X-rays and chalcopyrite atomic planes with distance d = 3.03 Å (space group I-42d, a = b = 5.290 Å and c = 10.426 Å, miller indexes h = 1, k = 1 and l = 2). With a background count of 1680 counts and a peak intensity of 572 counts, the chalcopyrite diffraction peak has a signal-to-noise ratio SNR ≈ 14.4, which passes the criteria of limit of detection $(SNR)_{LOD} = 3$ and limit of quantification $(SNR)_{LOQ} = 10$ [13]. One advantage of Rietveld refinement over classical straight line calibration method is the proper intensity extractions of overlapping peaks, such as chalcopyrite peak at 2θ = 34.2° and gypsum peak at 2θ = 34.0°, see Figure 3.

Similarly, the main diffraction peak of Bi-Sb-Pb-As sulphosalts partially overlap with chalcopyrite peak, Figure 4. The phase ID with HighScore Plus assigned the diffraction peak at 2θ = 34.74° to the tetrahedrite family, specifically to ICDD: 01-074-3633, a mercury-copper-antimony-arsenic sulfide. Such phase was selected as the best available structural model based on scores of search-match algorithm of HighScore, graphical refinement criteria and agreement with antimony elemental concentration in the concentrates, see Figure 7. From the net peak intensity I = 1425 counts and background intensity B = 3120 counts, the signal-to-noise ratio can be calculated: $SNR = I/\sqrt{B} \approx 25$.

3.3. Rietveld Refinements and Trends

The mineralogical quantification of the major phases is reported in Table 2. The chalcopyrite in the concentrate is 40–50 times higher than in the ore blends. Other major minerals which show higher percentage in the concentrate are sphalerite, tetrahedrite and pyrite.

The visualization of the Rietveld results of different ore blends over time allows the plants to assess and control the mineral variability of the feed used in the flotation cell. The most prominent difference in the ore blends is the type of gangue mineralogy, characterized by the amount of pyrite and chamosite, Figure 5. The flotation parameters, the surfactant selection and surfactant amount can therefore be optimized based on such quantities. The variation over time of the chalcopyrite content as well as other relevant minerals included in the Rietveld analysis is shown in Figure 6. Chalcopyrite is the main copper ore mineral and it varies between 0.5 wt% and 1.9 wt%, which is well above the limit of detection, see Figure 3. The limit of detection is approximately 120 net counts, as calculated from $LOD = [3 * net\ intensity * (background)^{-0.5}]$ [13]. A SNR > 10 is the minimum prerequisite to accurately quantify a mineral in a mixture. The accuracy of the Rietveld results depends also on (1) particles statistic, (2) degree of overlap with peaks from other phases, (3) preferential orientation, (4) overgrinding and (5) microabsorption.

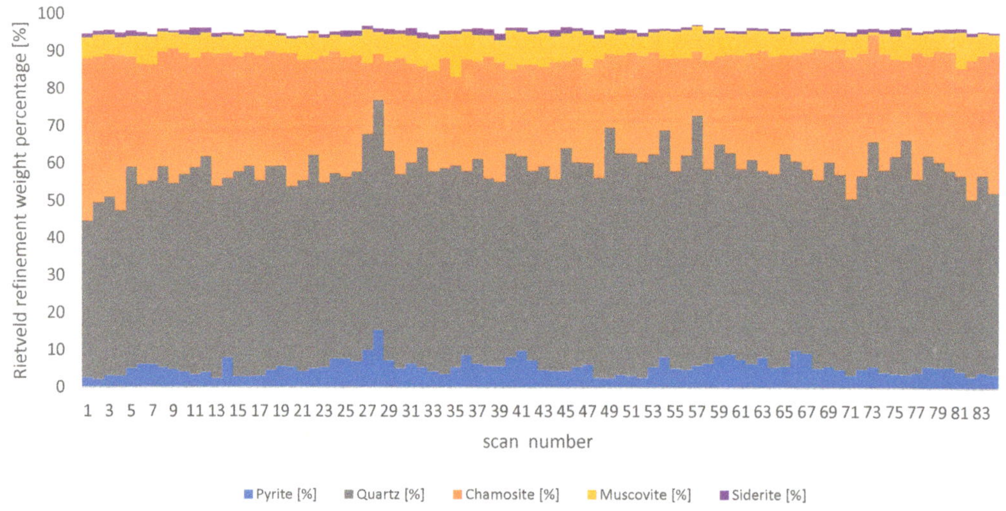

Figure 5. Mineralogical variability of ore blends for the major gangue minerals as determined by XRD and Rietveld refinement.

The accuracy of the XRD results was verified for all 265 materials by comparing the elemental composition calculated from the Rietveld refinements with the ICP values, see Figure 7. The elemental composition of copper, zinc and antimony can be fitted with straight lines which give, respectively, R^2 values of 0.99, 0.98 and 0.96, and slope values of 0.99, 1.05 and 0.94. The automatic Rietveld analysis gave accurate results with precision that depends on the mineral species and their concentrations. For instance, the precision of zinc calculated from Rietveld greatly decreases when the zinc content is below 0.2 wt%, which corresponds to a sphalerite concentration of 0.3 wt%, Figure 7. The relatively higher LOD of sphalerite is due to its strong overlap with pyrite diffraction peaks.

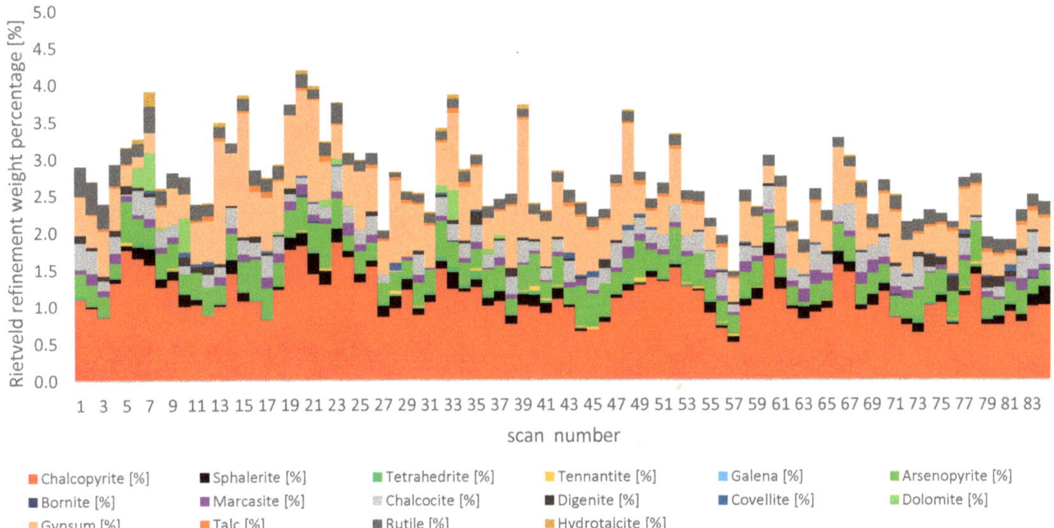

Figure 6. Mineralogical variability of ore blends for minor mineralogical phases, as determined by XRD and Rietveld refinement.

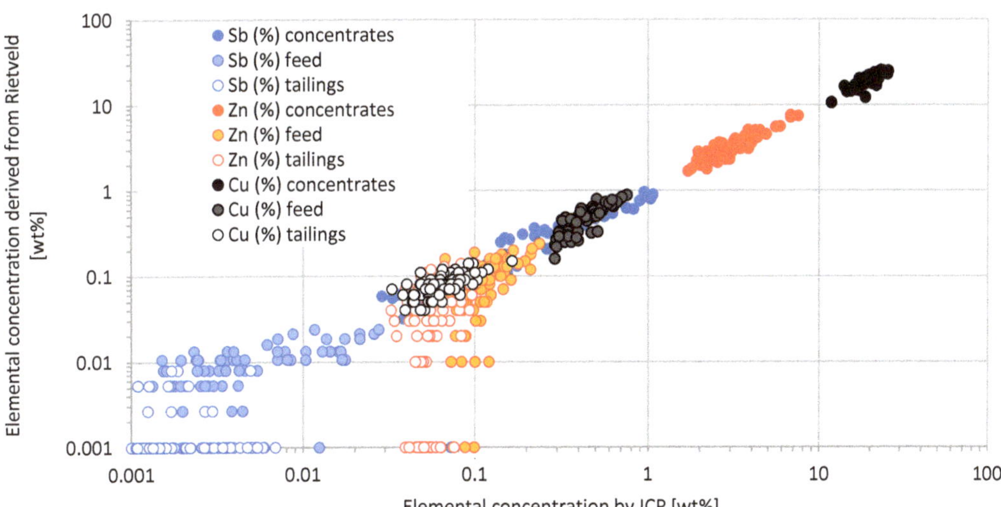

Figure 7. Agreement between chemical compositions obtained from ICP and Rietveld refinement. Zero values obtained by Rietveld are set to 0.001 for graphical purpose.

Deviation from expected results can be attributed to incorrect Rietveld structural models and refinement strategy, but most importantly, on sample preparation. The materials in this study are highly complex as they contain hard minerals, such as quartz and pyrite, and soft minerals, such as phyllosilicates, sphalerite and chalcopyrite. The mass absorption coefficient of pyrite is almost twice the value of quartz; therefore, microabsorption effects are likely to affect the analysis if samples are not properly ground. Vice versa, when samples are overly ground, the soft minerals can be partially amorphized. The disc mill used in this study has its disadvantages and advantages: it can affect sample crystallinity due to high energy impacts, but it fastens sample preparation operations, which can be integrated in a fully automated laboratory for quality control purposes. The use of dedicated XRD sample preparation procedures (milling and pressing parameter) and optimized data collection strategies can further improve the quantification of crystalline phases, including secondary copper minerals.

3.4. Statistical Methods—Principal Component Analysis (PCA): Clustering of Copper Concentrates

An alternative and powerful method to study the major minerals trends is the cluster analysis. In this example, the copper concentrate scans are grouped into clusters which are then visualized using Principal Component Analysis. During cluster analysis, the proprietary search-match algorithm of HighScore compares all the XRD scans and calculates a correlation matrix. The three principal components of the PCA substitute the 3300 datapoints of each XRD scan and they explain the majority of variance in the correlation matrix.

The first, second and third principal component explain, respectively, 50%, 31% and 9% of the variance, see the eigenvalue plot in Figure 8. In total, 89.55% of the variance of the correlation matrix is described in the PCA plot in Figure 9. The 90 different copper concentrates can be grouped into four major clusters plus an outlier that could not be grouped into any of the four clusters. The size of each marker is proportional to the pyrite content as determined by Rietveld refinement. The PCA plot shows that the pyrite content varies along the second principal component of the PCA plot. Each cluster has specific mineralogical properties; Table 3 focuses on the main differences among the clusters. The blue and gray cluster are the richest in chalcopyrite, but the blue cluster is higher in quartz and chamosite and lower in pyrite. The green and brown clusters are progressively richer in pyrite and poorer in chalcopyrite. Finally, the one outlier has a composition similar to the brown cluster, but it has a significant amount of secondary copper minerals, such as digenite Cu_9S_5 and covellite CuS.

Table 3. Average weight percentages of selected minerals of different clusters of copper concentrate, as determined by Rietveld refinements. The mineral variability within the same cluster is reported as the standard variation in between brackets. #Samples represents the number of samples included in each cluster.

Cluster Color	#Samples	Cluster Features	Minerals Quantification [wt%]			
			Chalcopyrite	Pyrite	Quartz	Chamosite
Blue Cluster	9	High $CuFeS_2$/Low FeS_2	60 (7)	12 (2)	10 (2)	6 (1)
Gray Cluster	53	High $CuFeS_2$/Medium FeS_2	63 (4)	20 (4)	4 (1)	2 (1)
Green Cluster.	25	Medium $CuFeS_2$/Medium FeS_2	50 (5)	33 (7)	4 (1)	1 (1)
Brown Cluster	2	Low $CuFeS_2$/High FeS_2	40 (13)	45 (14)	4 (1)	2 (1)
Black Outlier	1	Secondary copper minerals	35	45	3	1

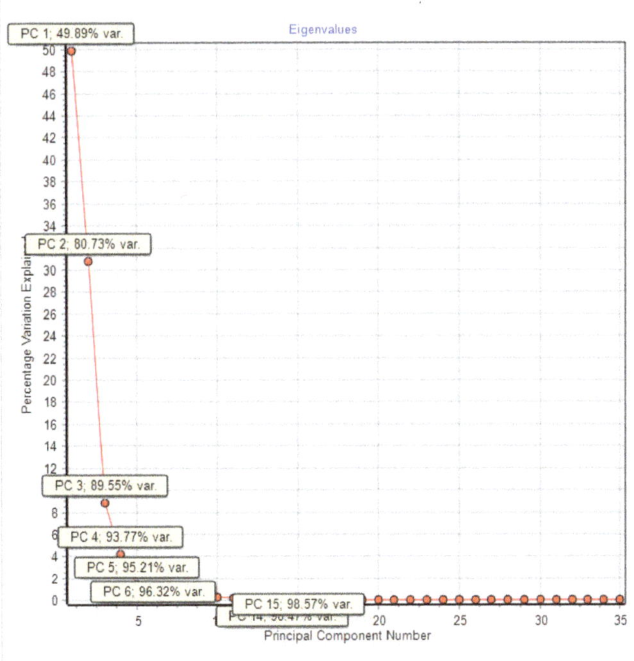

Figure 8. Eigenvalues plot explaining how the data variance is described by the most relevant principal components (PCs).

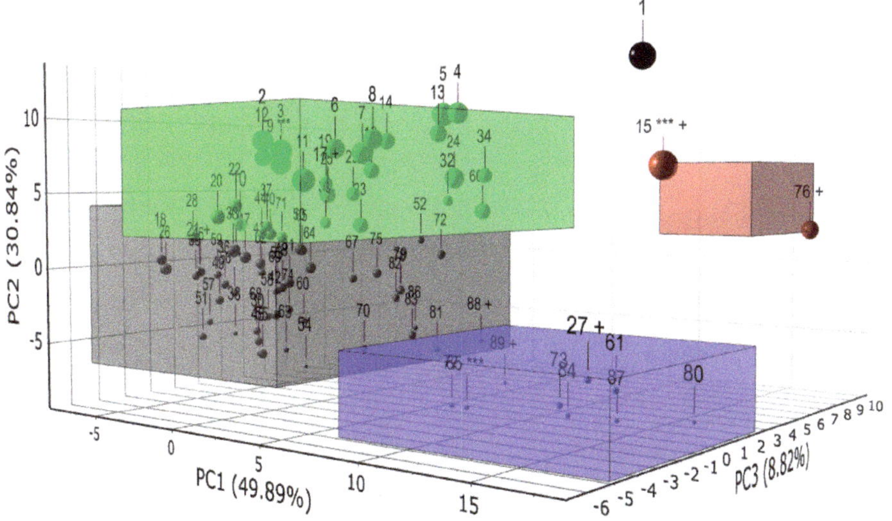

Figure 9. Principal Components Analysis plot of XRD scans of copper concentrates. Each scan is represented by a sphere with a radius proportional to its pyrite content. The four clusters and the one outlier are visualized in different colors. *** representative XRD scan of each cluster, + most different XRD scans within a cluster.

The XRD scans and the cluster algorithm parameter can be saved and used in the future to automatically classify an unknown material based on which cluster it belongs to. With such information, the processing plant can quickly and automatically decide if and where a copper concentrate needs to be further processed, blended and stored.

3.5. Statistical Methods—Partial Least Squares Regression (PLSR): Predicting Antimony in Copper Concentrates

Essential information for the mining process can also be extracted from the raw XRD scans by training the SIMPLS algorithm (statistically inspired modification of the partial least squares) [14] implemented in HighScore Plus. The algorithm belongs to the data mining method of partial least square regression. In this example, the training values of antimony content in the concentrates were provided by ICP chemical analysis. Since PLSR is a statistical method, several scans are needed to obtain a reliable PLSR calibration file. Here, we used 72 XRD scans for building the model and 17 XRD scans, i.e., 20% of the scans, to cross-calibrate the model (Figure 10). The latter were used to calculate the Root Mean Square Error of Prediction RMSEP = 556 ppm, which represents 1σ standard deviation of the predicted values. The PLSR model can then be used to estimate the antimony concentration of unknown samples without the need of Rietveld refinement, especially for copper concentrates with antimony concentration between 287 ppm and 10,650 ppm, the concentration range of the training values.

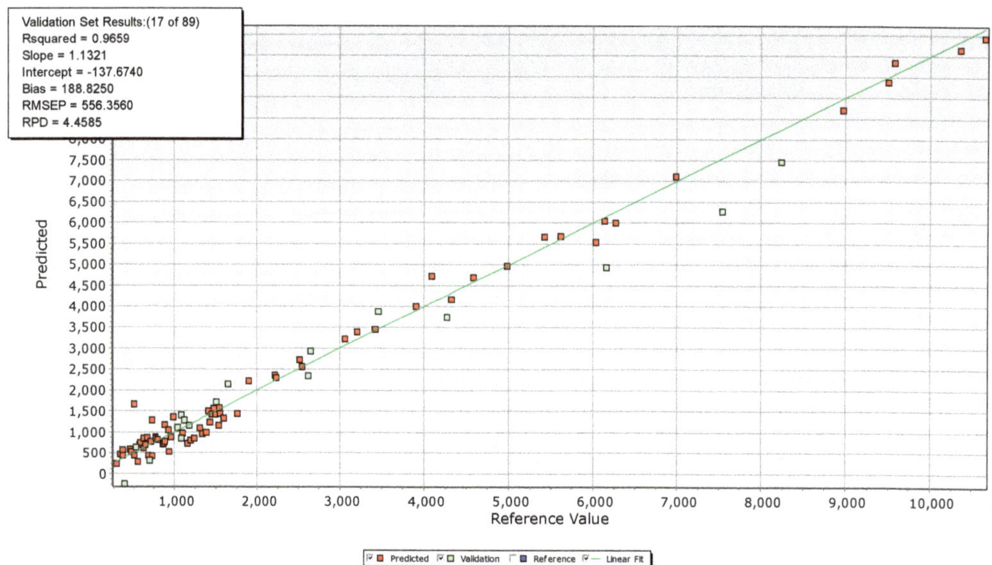

Figure 10. PLSR plot of predicted values and reference values of antimony concentration in copper concentrates. The plot contains both training and validation data of the model.

Several evaluation plots are available to assess the calibration model as well as extracting further information from the raw data. For instance, in Figure 11, the regression coefficients vs °2θ plot reveals the diffraction angles ranges most useful to build the PLSR calibration model. The regression coefficient has large values (>0.02) at the °2θ values of the tetrahedrite peaks, Table 4. As expected, this shows that the tetrahedrite is the main carrier of Sb in the analyzed samples. Such information was not accessible by wet chemistry or by Rietveld refinement alone and can be used to select the right surfactant to depress this mineral species. The comparison of the regression coefficient in Figure 11 with the relative peak intensity I[%] of tetrahedrite in Table 4 reveals how the PLSR model considers both the diffraction peak intensities as well as their degree of overlap with other phases. For instance, the less intense but isolated $2\theta = 28.2°$ peak, has a higher coefficient than the main peak at $2\theta = 34.7°$, partially overlapping with the chalcopyrite peak at $2\theta = 34.3°$ (Figure 4).

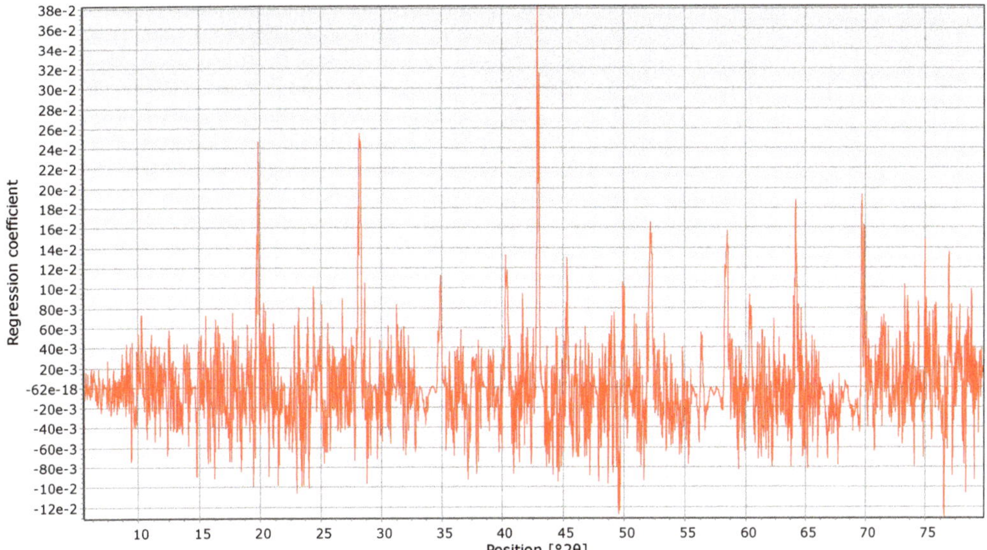

Figure 11. PLS regression coefficients for antimony concentration in concentrates. The coefficients identify the parts of the diffractogram contributing most to the PLSR regression model.

Table 4. 2theta angles in the diffraction pattern of copper concentrates with the highest regression coefficient for the PLSR model of antimony, and diffraction peaks of the corresponding phase.

$°2\theta$	Phase Assignment	H K L Miller Indexes	I [%] from 01-074-3633
18.9	Tetrahedrite	0 2 0	2.5
28.2	Tetrahedrite	0 2 2	7.1
34.7	Tetrahedrite	2 2 2	100
40.4	Tetrahedrite	0 4 0	21.2
42.9	Tetrahedrite	0 3 3	7.7
45.4	Tetrahedrite	0 4 2	1.6
50.0	Tetrahedrite	2 4 2	1.6
52.2	Tetrahedrite	1 4 3	6.1
58.4	Tetrahedrite	0 4 4	39.8
64.2	Tetrahedrite	1 6 1	6.3
69.7	Tetrahedrite	2 6 2	20.3

4. Discussion

In this study, we demonstrated the fast and accurate mineralogical monitoring of copper ore blends, tailing and concentrates. The fully automatable XRD analysis and Rietveld refinement of the Minerals Edition of Aeris provides both mineralogical overviews and precious details throughout the entire copper process, from exploration to final products. Such information is accessible to non-expert users and can be easily visualized on screen or stored into a LIMS system. Hard minerals, such as quartz, should be monitored to increase the lifetime of crushing and milling equipment. The type of copper bearing minerals (such as chalcopyrite, digenite and covellite) allows a better selection of collectors in the flotation process. Similarly, the correct depressant and suspension chemistry should be selected to suppress the flotation of a minerals bearing penalty elements, such as tetrahedrite, which contains antimony.

Accurate quantitative mineralogical analysis is not the only added value of XRD. If the minerals carry information of specific properties of a material, the information can also be extracted using algorithms other than Rietveld. Clustering and predictive tools, such as PCA and PLSR, add further possibilities to process control and allow mineral processing engineers and scientist to collect invaluable insights. In this work, these two analysis techniques were evaluated and compared with classical Rietveld results. They can be even easier to use than crystallographic information and Rietveld experience is not needed.

PCA analysis is particularly useful for qualitative or discriminatory analysis, as well as to simplify the classification problem of complex high dimensional data. As the higher-grade deposits are depleted, there is an increasing need to beneficiate more challenging ore bodies. XRD and PCA can be used to mix different ore grades to an optimal blend, adapt the downstream process and increase recovery [15]. Cluster analysis can be used to identify samples which best describe groups of materials and it reduces the time spent by staff performing manual analysis. The representative scans of each cluster as well as the most diverse scans within each cluster are automatically calculated. PCA can also quickly identify outliers: samples with peculiar mineralogical composition or badly prepared and mislabeled specimens. Therefore, PCA can fasten and improve the process of data cleaning before further processing the data.

Thanks to the modern detectors and diffractometers, it is easy and cheap to collect tens of training sets to build PLSR calibration models. The implementation of such models on the Aeris extracts indirect properties from the raw XRD data, such as their chemistry or more complex quantities, such as process-relevant parameters [16]. In general, any physical or chemical quantity with a high degree of correlation with the mineralogical composition of the material can be estimated using PLSR. Therefore, XRD and PLSR can replace more time- and cost-consuming analytical techniques, such as wet chemistry.

In conclusion, X-ray diffraction (XRD) is an essential tool for mineralogical analysis and it can be easily implemented in process environments to improve mine operations and save costs.

Author Contributions: Conceptualization, M.P., Á.L., D.D., U.K. and N.N.; Data curation, M.P. and M.M.; Formal analysis, M.P. and N.N.; Investigation, M.P., D.D. and M.M.; Methodology, M.P. and M.M.; Project administration, D.D.; Resources, Á.L. and U.K.; Supervision, D.D. and U.K.; Validation, M.M.; Writing—original draft, M.P.; Writing—review & editing, D.D., U.K. and N.N. All authors have read and agreed to the published version of the manuscript.

Funding: This research received no external funding.

Data Availability Statement: All data were gathered at Atalaya Mine Plc. laboratories at Minas de Riotinto (Spain) and treated at Malvern Panalytical B.V laboratories in Almelo (The Netherlands).

Acknowledgments: The support of Solvay, Atalaya Mining and Malvern Panalytical management for allowing this project is gratefully acknowledged.

Conflicts of Interest: The authors declare no conflict of interest.

References

1. Gleisberg, D. *Mineral Commodity Summaries*; Government Printing Office: District of Columbia, WA, USA, 2021; Volume 3. [CrossRef]
2. Leja, J. *Surface Chemistry of Froth Flotation*; Springer: Boston, MA, USA, 1981. [CrossRef]
3. Rietveld, H.M. Line Profiles of Neutron Powder-Diffraction Peaks for Structure Refinement. *Acta Crystallogr.* **1967**, *22*, 151–152. [CrossRef]
4. Paine, M.; König, U.; Staples, E. Application of Rapid X-Ray Diffraction (Xrd) and Cluster Analysis to Grade Control of Iron Ores. In *Proceedings of the 10th International Congress for Applied Mineralogy (ICAM)*; Springer: Berlin/Heidelberg, Germany, 2012; pp. 495–501. [CrossRef]
5. König, U.; Norberg, N.; Gobbo, L. From Iron Ore to Iron Sinter—Process Control Using X-Ray Diffraction (XRD). In *Anais dos Seminários de Redução, Minério de Ferro e Aglomeração*; Editora Blucher: São Paulo, Brazil, 2017; pp. 146–153. [CrossRef]

6. König, U.; Degen, T.; Norberg, N. PLSR as a New XRD Method for Downstream Processing of Ores:—Case Study: Fe2+ Determination in Iron Ore Sinter. In *Powder Diffraction*; Cambridge University Press: Cambridge, UK, 2014; Volume 29, pp. S78–S83. [CrossRef]
7. König, U.; Gobbo, L.; Reiss, C. Quantitative XRD for Ore, Sinter, and Slag Characterization in the Steel Industry. In *Proceedings of the 10th International Congress for Applied Mineralogy (ICAM)*; Springer: Berlin/Heidelberg, Germany, 2012; pp. 385–393. [CrossRef]
8. Galluccio, S.; Pöllmann, H. Quantifications of Cements Composed of OPC, Calcined Clay, Pozzolanes and Limestone. In *RILEM Bookseries*; Springer: Berlin/Heidelberg, Germany, 2020; Volume 25, pp. 425–442. [CrossRef]
9. Tornos, F. Environment of Formation and Styles of Volcanogenic Massive Sulfides: The Iberian Pyrite Belt. *Ore Geol. Rev.* **2006**, *28*, 259–307. [CrossRef]
10. Degen, T.; Sadki, M.; Bron, E.; König, U.; Nénert, G. The High Score Suite. In *Powder Diffraction*; Cambridge University Press: Cambridge, UK, 2014; Volume 29, pp. S13–S18. [CrossRef]
11. Gates-Rector, S.; Blanton, T. The Powder Diffraction File: A Quality Materials Characterization Database. *Powder Diffr.* **2019**, *34*, 352–360. [CrossRef]
12. Rietveld, H.M. A Profile Refinement Method for Nuclear and Magnetic Structures. *J. Appl. Crystallogr.* **1969**, *2*, 65–71. [CrossRef]
13. Shrivastava, A.; Gupta, V. Methods for the Determination of Limit of Detection and Limit of Quantitation of the Analytical Methods. *Chronicles Young Sci.* **2011**, *2*, 21. [CrossRef]
14. De Jong, S. SIMPLS: An Alternative Approach to Partial Least Squares Regression. *Chemom. Intell. Lab. Syst.* **1993**, *18*, 251–263. [CrossRef]
15. Lance, G.N.; Williams, W.T. A General Theory of Classificatory Sorting Strategies: 1. Hierarchical Systems. *Comput. J.* **1967**, *9*, 373–380. [CrossRef]
16. König, U. Process Monitoring for Iron Ore Pelletizing—XRD in Combination with PLSR. In *Iron Ore*; AUSIMM: Perth, Australia, 2019; pp. 121–126.

minerals

MDPI

Article

Effects of Pyrite Texture on Flotation Performance of Copper Sulfide Ores

İlkay B. Can *, Seda Özçelik and Zafir Ekmekçi

Department of Mining Engineering, Hacettepe University, Beytepe, Ankara 06800, Turkey;
sedaozcelik26@gmail.com (S.Ö.); e.zafir@gmail.com (Z.E.)
* Correspondence: ilkay@hacettepe.edu.tr

Abstract: Pyrite particles, having framboidal/altered texture, are known to significantly affect pulp chemistry and adversely affect flotation performance. Therefore, the main objectives of this study were to demonstrate influence of pyrite mineralogy on the flotation of copper (sulphidic) ores and develop alternative conditions to improve the performance. Two copper ore samples (Ore A and Ore B) having different textural/modal mineralogy and flotation characteristics were taken from different zones of the same ore deposit. Ore B contained framboidal pyrite and altered pyrite/marcasite, which is considered the main reason for the low flotation performance in both copper and pyrite flotation sections of the process plant. Flotation tests were conducted under different conditions using the two ore samples and a 50:50 blend. The results showed that Ore A could be concentrated under the base conditions, as applied in the existing flotation plant. On the other hand, Ore B did not respond to the base conditions and a copper recovery of only 5% could be obtained. Besides, blending Ore B with Ore A negatively affected the flotation behavior of Ore A. An alternative flotation chemistry was applied on Ore B using Na_2S for surface cleaning and Na-Metabisulfite (MBS) for pyrite depression in the copper flotation stage. The surface cleaning reduced the rate of oxidation of the framboidal pyrite in Ore B. As a result, the copper recovery could be increased to 52% Cu for Ore B, and 65% for the mixed ore sample.

Keywords: framboidal pyrite; sulfide minerals; flotation; process mineralogy

Citation: Can, İ.B.; Özçelik, S.;
Ekmekçi, Z. Effects of Pyrite Texture
on Flotation Performance of Copper
Sulfide Ores. *Minerals* **2021**, *11*, 1218.
https://doi.org/10.3390/min11111218

Academic Editors: Herbert Pöllmann
and Uwe König

Received: 18 October 2021
Accepted: 28 October 2021
Published: 1 November 2021

Publisher's Note: MDPI stays neutral
with regard to jurisdictional claims in
published maps and institutional affiliations.

1. Introduction

Pyrite (FeS_2) is the most abundant sulfide mineral and usually occurs both as a primary and secondary mineral according to its genesis. Lattice substitution of some minor and trace elements such as nickel, cobalt, arsenic, lead, and gold are also the results of this ore genesis and geographical location [1] and may have an impact on pyrite floatability [2]. Pyrite is widespread in hydrothermal veins, modern and ancient sedimentary (volcano, exhalative) rocks, contact metamorphic deposits/rocks and as an accessory mineral in igneous rocks [3]. The genesis of pyrite has been considered as one of the main causes of variation in surface chemical characteristics and, thereby, of the differences in their floatability [4]. Contact angle measurements of different types of pyrite showed a significant relationship between the origin of pyrite and its wetting characteristics. Flotation characteristics of pyrite are also influenced by morphology, crystallography, and the presence of impurities in the crystal structure [5]. The marcasite mineral (FeS_2), the polymorph of pyrite, has different crystallography and stability and, hence, different flotation behaviour than that of pyrite, even when they exist in same deposits [3,6].

Pyrite has a simple cubic crystal structure, while marcasite, which has the same chemical composition as pyrite, has an orthorhombic crystal structure [7]. Another micro-morphological feature, called framboid, is also described for pyrite. Framboidal form of pyrite can be defined as raspberry-shaped masses. This spherical structure is composed of numerous microcrystals, which are equant and equidimensional [8]. This microcrystal packing is typically irregular and disordered. A framboidal morphology of pyrite is one of the common ore textures observed in disseminated massive sulfide ore deposits [9]. It is also ubiquitous in modern sediments and ancient sedimentary rocks but occurs less frequently in hydrothermal deposits [10].

Oxidation of pyrite produces hydroxide/oxide coating products which have significant effects on floatability and associated minerals. The rate of surface oxidation varies as a function of pulp chemistry, the grinding conditions and mineralogical characteristics. For example, marcasite (FeS_2) is more reactive than pyrite and reacts more readily with moisture and oxygen [11,12]. Relative reaction rates for different morphological forms of pyrite are described in the order of marcasite > framboidal pyrite > massive pyrite [13]. The stabilities of these minerals are different, and framboidal pyrite decomposes more readily under mildly oxidizing conditions than cubic pyrite [14]. Thus, it can be said that different crystalline pyrite forms would react differently to mild oxidation conditions during flotation. Miller et al. [15] observed the same finding for reactive auriferous pyrite. Air was replaced by nitrogen gas to remove the dissolved oxygen from the flotation system and thereby auriferous pyrite oxidation could be minimized [16], particularly for framboidal pyrite and marcasite.

It is known that selective flotation of the base metal sulfide minerals from pyrite is strongly influenced by the type and concentration of metal oxidation species produced during grinding [17]. These species may affect the floatability of minerals, rendering the surfaces hydrophilic or leading to inadvertent activation [18,19]. The pyrite content of the flotation feed also plays an important role in increasing the oxidation of minerals as it consumes dissolved oxygen in the flotation pulp [20]. Since the framboidal pyrite has a large surface area, the oxidation rate is faster than with the standard texture of pyrite, producing a variety of iron oxidation products. It has been proposed that the presence of framboidal pyrite in an ore is a critical parameter affecting Pb flotation rather than the high amount of purely pyrite, since it increases the rate of galvanic interaction with other metal sulfides [5].

Pyrite is the most common of the gold-bearing minerals and, therefore, pyrite morphology is of paramount importance for its association with gold. Submicroscopic association with framboidal pyrite is one of the main textures observed in gold deposits [21]. Simmons [14] reported that the occurrence of gold in auriferous pyrite may differ, as relatively low levels of gold were detected in coarse grained pyrites, whereas fine grained, amorphous and framboidal pyrite contained much higher levels of gold. In other words, the framboidal structure of pyrite may have the highest intrinsic gold value. However, significant amounts of gold loss to tailing may occur due to the problems in floatability of fine grained, framboidal pyrite particles.

Similar effects are also observed in leaching gold from pyrite particles. The pyrite source is an important parameter in the kinetic behaviour of bacterial oxidation of pyrite, particularly for gold recovery from refractory types of ores. It is reported that the oxidation of framboidal and euhedral pyrites are completely different. Framboidal pyrite has a granular and irregular surface structure and is more chemically reactive than the highly crystalline surface structure of euhedral pyrite. Hence, the rate of surface oxidation by *Thiobacillus ferrooxidans* microorganism is higher than the other pyrite forms [22].

Previous works in the literature have shown that the textural mineralogy of pyrite determines its surface characteristics and flotation behaviour in complex sulfide ores. Pyrite particles having a framboidal/altered texture are known to significantly affect pulp chemistry and adversely affect flotation performance. Therefore, the main objectives of this study were to demonstrate the influence of pyrite mineralogy on the flotation of copper ores, and to develop an alternative condition to improve performance. Two copper ore samples (Ore A and Ore B) having different textural/modal mineralogy and flotation characteristics were taken from different zones of the same ore deposit. Ore B contained framboidal pyrite and altered pyrite/marcasite, which is considered the main reason for the low flotation performance in both copper and pyrite flotation sections of the process plant.

2. Materials and Methods

Representative samples were taken from the two ore types, named as Ore A and Ore B, from a complex Cu-Zn ore deposited located in northeast of Turkey. Both ore types were treated by flotation at a primary grind size of 80% passing 38 µm to produce copper and pyrite concentrates. Production of a separate zinc concentrate was not considered feasible in the operation due to the highly variable zinc grade of the ore. Therefore, the zinc was recovered to the copper concentrate. Table 1 shows size-by-size assays of different major metals in the two ore samples ground to d_{80}: 38 µm. Ore A was a typical copper ore containing high amount of pyrite. On the other hand, Ore B contained significant amounts of Zn and Pb, indicating its complex nature.

Table 1. Chemical composition of Ore A and Ore B on a size by assay basis.

		Weight, %	Cu, %	Zn, %	Pb, %	Fe, %	S, %
	+38 µm	22.30	2.09	0.10	0.04	45.44	51.24
	−38 + 20 µm	25.90	2.39	0.10	0.07	45.75	50.63
Ore A	−20 + 10 µm	49.02	2.99	0.19	0.10	44.98	47.20
	−10 µm	2.78	3.45	0.38	0.27	40.10	39.90
	Head assays	100.00	2.94	0.20	0.08	44.47	50.31
	+38 µm	20.89	0.84	2.75	0.60	33.27	38.54
	−38 + 20 µm	21.86	0.82	2.71	0.83	33.04	35.36
Ore B	−20 + 10 µm	51.26	1.60	3.97	1.20	26.94	31.20
	−10 µm	5.99	1.88	4.43	1.35	23.56	27.40
	Head assays	100.00	1.34	3.66	1.08	28.03	33.90

BMA (Bulk Mineralogical Analysis) and PMA (Particle Mineral Analysis) characterization were performed on a size-by-size basis for both samples. Polished sections of +38 µm, −38 + 20 µm, and −20 + 10 µm size fractions were prepared and analyzed using QemSCAN, which has an FEI Quanta 650F electron microscope equipped with a field-emission gun as an electron source. The −10µm size fraction was not included in the mineralogical analysis because of the agglomeration of fine particles observed during preparation of the polished sections. For imaging and X-ray based microchemical analysis, the instrument is equipped with a four-quadrant, solid state back-scattered electron (BSE) detector and two Bruker XFlash 5030 detectors. The liberation criterion was selected as 90%, which means any particle containing 90% and 100% by area of the mineral of interest is regarded as liberated. Particles that contain between 50% and 90% by area are referred to as middling/binary particles, while those containing less than 50% by area are referred to as locked particles.

In addition to quantitative mineralogical characterization, petrography analysis by transmitted and reflective light microscopy was performed to determine Fe sulfide speciation. The samples were prepared as 20×40 mm polished thin sections and analyzed with a petrographic microscope under polarized transmitted and polarized reflected light. This analysis provides information about petrographic rock classification, microstructure of the samples, and the modal percentage of each mineral.

Ethylene diamine-tetra acetic acid disodium salt (EDTA) extraction tests were conducted to characterize surface oxidation of the sulfide minerals in both ore samples. A fraction (10 g) of dry ore was leached for 30 min in a 200 mL solution containing 3% EDTA at pH 7.5. The pulp was filtrated, and the filtrate was assayed for Cu, Zn, Fe, Pb using AAS. The results indicated that surface oxidation of the copper minerals and galena was higher in Ore B (Table 2).

Batch scale flotation tests were performed on both ore types and on a composite sample prepared from their blend to evaluate the effects of Fe sulfide forms on flotation. The flotation feed was ground to d_{80}: 38 µm in a ball mill at 60% w/w pulp density. The flotation tests were carried out at 30% pulp density using a 4.5 L Denver flotation cell. In the flotation plant, a proprietary blend collector Kimfloat900 (150 g/t), a thionocarbamate formulation, was used in the copper flotation circuit at pH 11.5–12. The pH was kept constant at the target values by adding lime into the ball mill and flotation. A TPS meter was used to measure pH and Eh values during the flotation process. Following the copper flotation stage, an ether amine form collector, TomAmine (100 g/t), was used as the collector for pyrite flotation at pH 11. These flotation conditions were termed the Base Condition (BC) which is applied in the plant. Alternative collectors, diisobutyl phosphinate (Aerophine 3418A), sodium isopropyl xanthate (SIPX) and thionocarbamate (Aero 5100) were tested to improve copper flotation performance. Potassium amyl xanthate (KAX) was tested for pyrite flotation. Methyl isobutyl carbinol (MIBC) was used as the frother. Sodium sulfide (Na_2S) and sodium hydrosulfide (NaHS) were tested for surface cleaning of the tarnished minerals and framboidal pyrite, particularly for Ore B (Table 3).

Table 2. EDTA results for Ore A and B.

	Extractable Metal/Total Metal (%)			
	Cu	**Fe**	**Pb**	**Zn**
Ore A	1.67	0.11	44.90	4.51
Ore B	3.69	0.09	72.70	0.86

Table 3. Reagent scheme used in the flotation tests.

Collectors for copper flotation	Kimfloat900 (used as base condition, BC) Aero5100 Aerophine 3418A SIPX
Collectors for pyrite flotation	TomAmine (used as base condition, BC) KAX
Frother	MIBC
Sulphidization agents for surface cleaning	NaHS, Na_2S
Depressant	Na-MBS

3. Results and Discussion

3.1. Modal/Particle Mineralogy and Fe Sulfide Identification

Mineral distribution is given on a size-by-size basis for both ore samples in Table 4. The major sulfide minerals were pyrite/marcasite, chalcopyrite, sphalerite, and minor amount of galena. The Ore B sample contained significant amounts of barite as the non-sulfide gangue mineral. Pyrite/marcasite accounted for the major iron sulfide gang mineral together with the altered pyrite in both ore types.

Grain size and liberation characteristics of the sulfide minerals were completely different in the two ore samples. Both chalcopyrite and pyrite showed higher liberation in Ore A with values of more than 80% and 90%, respectively (Table 5). Percentage of free particles decreased at −20 + 10 μm size fraction in both ore samples. This unexpected result was attributed to agglomeration problems during preparation of polished sections. Unlike Ore A, the sulfide minerals in Ore B were in the form of fine grains (Figure 1) and showed complex liberation characteristics. It was seen that in every size fraction, the grain size distribution was finer in Ore B, particularly for pyrite (Figure 1). The liberation value of chalcopyrite in Ore B was only ~15%, and the chalcopyrite particles were mostly in the form of binary chalcopyrite/pyrite particles. The percentage of sphalerite was higher in Ore B and had a low degree of liberation, in the range of 18–33%. A significant part of sphalerite was in the form of binary association with pyrite and barite. The pyrite particles in Ore A showed a considerably higher degree of liberation (91%) than those in Ore B (57%). The aggregate and inclusion-rich form of pyrite was also seen from binary association data. The back-scatter electron (BSE) images clearly showed that some of the pyrite in Ore B had a framboidal structure and the rest had an altered structure and marcasite (Figure 2).

Table 4. Modal mineralogy of Ore A and Ore B on size-by-size basis.

Minerals	Ore A (%)			Ore B (%)		
	+38 μm	−38 + 20 μm	−20 + 10 μm	+38 μm	−38 + 20 μm	−20 + 10 μm
Chalcopyrite	5.96	6.54	7.33	1.29	1.16	1.68
Sphalerite	0.08	0.07	0.19	3.99	3.96	4.35
Pyrite/Marcasite	90.72	89.59	85.52	66.83	56.71	40.75
Galena	<0.01	<0.01	<0.01	0.67	0.94	0.37
Barite	0.05	0.04	0.10	22.17	25.62	18.06
Quartz	0.03	0.06	0.07	0.02	0.07	0.03
Biotite	0.07	0.01	<0.01	0.09	0.03	<0.01
Pyrite-Altered/Aggregate	2.85	3.45	6.43	2.59	7.46	17.47
Sulfide-Clay Mixed	0.09	0.04	0.05	0.03	0.07	0.21
Sulfides-Barite Aggregates	0.01	0.01	0.04	2.13	3.64	16.14
Others	0.13	0.18	0.27	0.19	0.34	0.94
Total	100.00	100.00	100.00	100.00	100.00	100.00

Table 5. Size-by-size liberation of chalcopyrite and pyrite minerals in Ore A and Ore B.

Chalcopyrite	Free	Binary Association							Total
		Pyrite	Sphalerite	Barite	Galena	Quartz	Aggregates	Other	
Ore A/+38 μm	80.53	17.80	0.21	0.17	0.00	0.00	0.07	1.23	100.00
Ore A/−38 + 20 μm	82.81	15.63	0.17	0.23	0.00	0.00	0.19	0.97	100.00
Ore A/−20 + 10 μm	56.21	38.05	0.25	0.27	0.01	0.00	3.96	1.25	100.00
Ore B/+38 μm	5.72	50.37	7.52	7.34	0.21	0.00	0.08	28.76	100.00
Ore B/−38 + 20 μm	15.13	39.33	8.18	16.62	0.11	0.00	0.82	19.82	100.00
Ore B/−20 + 10 μm	13.89	34.49	6.56	9.28	0.08	0.06	7.00	28.63	100.00

Pyrite	Free	Binary Association							Total
		Sphalerite	Barite	Galena	Chalcopyite	Quartz	Aggregates	Other	
Ore A/+38 μm	92.05	0.43	0.06	0.01	5.28	0.00	1.99	0.18	100.00
Ore A/−38 + 20 μm	91.18	0.40	0.03	0.00	5.25	0.01	2.79	0.34	100.00
Ore A/−20 + 10 μm	81.74	0.49	0.25	0.01	11.59	0.00	5.14	0.77	100.00
Ore B/+38 μm	77.01	10.67	7.30	0.18	2.07	0.00	2.03	0.75	100.00
Ore B/−38 + 20 μm	51.09	7.55	6.45	0.10	2.27	0.00	31.44	1.09	100.00
Ore B/−20 + 10 μm	44.39	8.48	14.37	0.24	2.94	0.01	15.87	13.71	100.00

The framboidal structure is very well defined by Wilkin and Barnes [10], and is considered to form as a result of consecutive processes such as nucleation and the growth of initial iron monosulfide microcrystals, reaction of the microcrystals to greigite, framboid growth of microcrystals and replacement of these framboids by pyrite. Since it is very difficult to quantify Fe sulfide species with QemSCAN, petrography analysis was performed using a transmitted and reflective light microscope on both ore types. According to the analysis of thin sections of Ore B, ~1.5% of pyrite particles were defined as framboidal with a size range of 0.01–5 mm. The marcasite particles were observed as alteration and weathering minerals in an amount of 2–3%, with a size range up to 0.3 mm. In some fragments, the pyrite crystals were immersed within a second generation of pyrite. In other fragments, the pyrite was intergrown with sphalerite (Figure 3a), and the two minerals formed crustiform intergrowths. Although only 1.5% of the pyrite appeared in framboidal form, the rest of the pyrite was found to exist with spongy inclusions, forming aggregates and an anhedral crystal form (Figure 3b). These aggregates tended to form rounded framboidal aggregates, which were clearly distinguished by the subhedral and inclusion-free crystals or the inclusion-poor interstitial aggregates of pyrite.

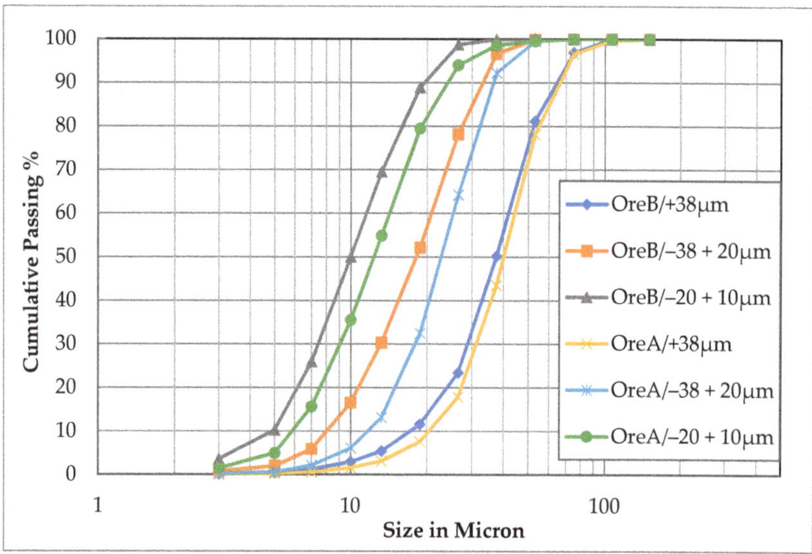

Figure 1. Grain size distributions of pyrite in size fractions of Ore A and B.

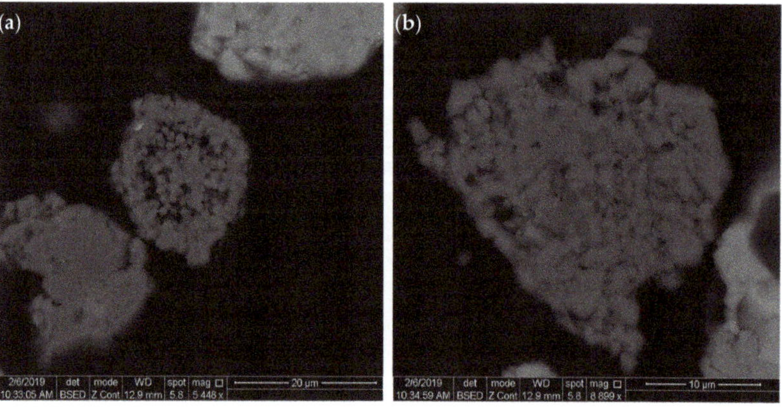

Figure 2. Framboidal pyrite (**a**) and altered pyrite/marcasite (**b**) from Ore B.

In Ore A, pyrite dominated the composition of the fragments and formed quasi-massive aggregates intergrown with subordinate crystals of chalcopyrite and marcasite (Figure 4). Pyrite accounted for 95–97% of modal mineralogy ranging in a size of 0.01 mm to massive. The amount of marcasite was 1.5% and found as fine-to medium-grained anhedral crystals which were heterogeneously dispersed in pyrite.

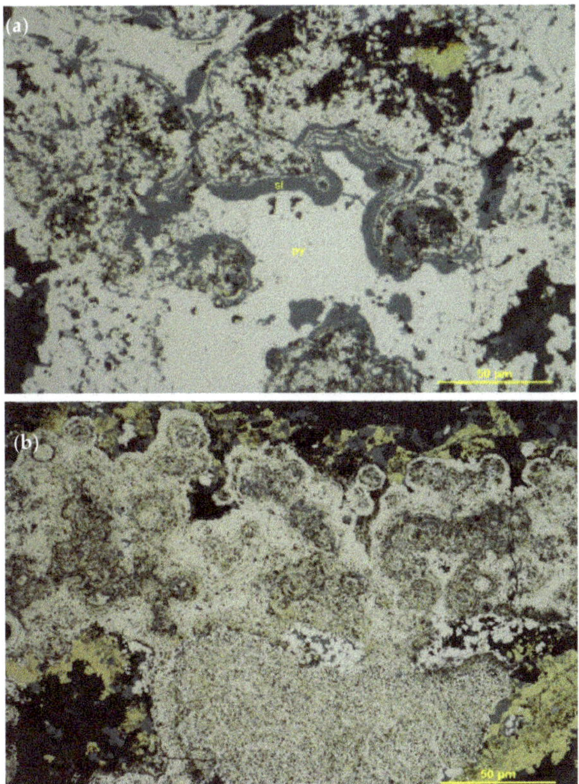

Figure 3. Very fine-grained crystals of pyrite (py) are intergrown with the sphalerite (sl) and define crustiform and layered intergrowths (**a**). Porous and spongy aggregates of pyrite define rounded framboidal aggregates and irregularly shaped domains associated with subordinate inclusion-free interstitial pyrite (**b**) from Ore B.

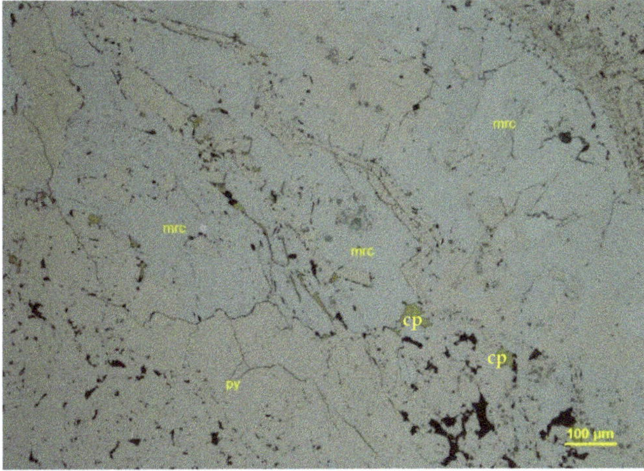

Figure 4. Pyrite (py) hosts medium-grained crystals of marcasite (mrc) and fine-grained crystals of chalcopyrite (cp) from Ore A.

3.2. Flotation Behavior of Ore A

Mineralogical investigation and chemical assays showed that Ore A had a relatively simpler ore texture and low Zn content. The test performed at the Base Condition (BC) produced a copper rougher concentrate at 71% copper recovery (Table 6). Pyrite flotation was applied on the copper rougher tail using TomAmine at pH 10.5–11. A pyrite concentrate was produced assaying 50% S at 68% recovery.

Table 6. Flotation behavior of Ore A under the Base Condition.

| Cu Rougher Concentrate | Mass Pull, % | Cu Rougher Concentrate | | | | Pyrite Concentrate | |
| | | Grade (%) | | Recovery (%) | | Grade (%) | Recovery (%) |
		Cu	Zn	Cu	Zn	S	S
Base Condition	13.69	13.99	0.74	70.98	48.93	50.31	68.22

3.3. Flotation Behavior of Ore B

Chemical and mineralogical characteristics of Ore B were completely different from those of Ore A. Rougher kinetic tests were conducted to determine its flotation response at the Base Condition and investigate effects of alternative flotation chemistry (collector type, sulphidization for surface cleaning, use of MBS as depressant) and particle size. In the BC test, stage addition of 150 g/t Kimfloat900 collector was applied at about pH 12. A combined copper rougher concentrate was produced assaying 2.4% Cu at 5.1% recovery. Following the copper flotation stage, pyrite rougher flotation was performed at the same pH using 100 g/t TomAmine as collector. A pyrite rougher concentrate was produced assaying 28% S at 25% recovery and 29% mass pull. These results show that the performance of both copper and pyrite flotation stages were considerably lower than those obtained with Ore A (Table 6).

Alternative flotation conditions were investigated to improve the flotation performance of Ore B. In one of the tests, the collector was added at the milling stage. Aero5100 (120 g/t) and a mixture of Aero3418A + SIPX (90 + 90 g/t) were used as alternative collectors. Figure 5 shows that the copper recovery increased to 67% at 31% mass pull using a mixture of Aero3418A + SIPX.

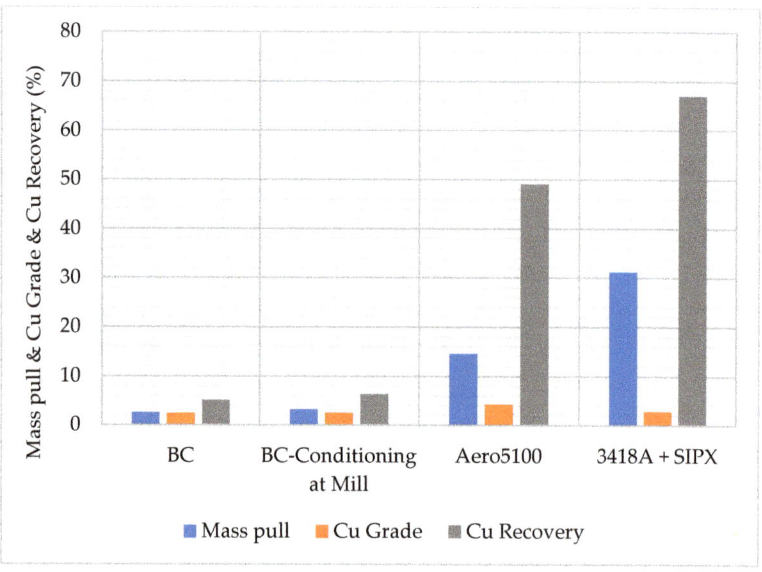

Figure 5. Effect of different reagent applications on the rougher flotation performance of Ore B.

3.3.1. Effects of Surface Cleaning

Ore B contained altered marcasite and framboidal pyrite, which oxidizes faster than pyrite. Therefore, Na_2S and NaHS were tested as sulphidization agents in order to remove the surface oxidation species from mineral surfaces and minimize further oxidation during flotation [9,23]. The mixture of Aerophine 3418A + SIPX was used as a collector in these tests at natural pH (pH 8.8–9.6). Na_2S was tested in 500 g/t, 1000 g/t and 1500 g/t dosages and added at the grinding stage. Addition of Na_2S decreased the pulp Eh as a function of reagent dosage down to the lowest value −150 mV (Ag/AgCl) after grinding. Pre-aeration (5 min) was applied prior to collector addition to increase the dissolved oxygen content and Eh to slightly positive values (−2 to 60 mV).

The results showed that sulfidization improved the flotation performance and the Cu recovery increased to 75% at a 1000 g/t Na_2S dosage. Increasing Na_2S dosage to 1500 g/t did not further improve the copper recovery (Figure 6). NaHS was also tested as an alternative sulfidizing reagent to Na_2S at a 1000 g/t dosage. Similar recovery values were obtained using both Na_2S and NaHS, and a 1000 g/t dosage was required for an effective surface cleaning.

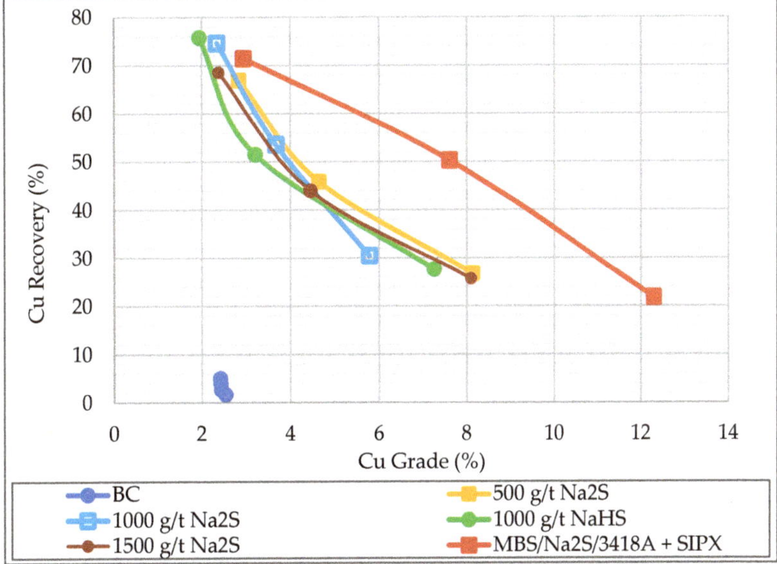

Figure 6. Effect of sulphidization and MBS addition on Cu recovery-grade of Ore B.

3.3.2. Effects of MBS Addition

Effective depression of pyrite during the copper rougher flotation stage directly affects the quality of both the copper rougher concentrate and the pyrite concentrate. It is well known that sulfoxy reagents such as sodium metabisulfite are widely used as for pyrite depression as they enhance formation of hydrated iron oxide layers on pyrite and suppress the adsorption of xanthate by reducing the mixed potential [19,24]. For this purpose, Na-metabisulfite (MBS) addition in the grinding stage was tested. The pulp potential (Eh) was monitored and recorded as −198 mV and 40 mV just after the grinding and pre-aeration stages, respectively. It was found that pyrite could be effectively depressed at 3000 g/t dosage and pH 6.5. Figure 6 shows that the highest copper grade values were obtained in the presence of MBS because of the effective depression of pyrite.

3.4. Flotation Behavior of Mix Ore

During plant scale operation, it may not be possible to always process Ore B and Ore A separately. A blend of these two ore types could be treated in the flotation plant. Therefore, effects of blending Ore B with Ore A were investigated under two different flotation conditions. The two ore types were blended at equal proportions for these tests according to the recommendations from the mine site.

The negative effect of Ore B on flotation performance was clearly seen in the mixture. The copper recovery in Ore A was 71% under the base flotation conditions (Table 6), and dropped to 22% when it was mixed with Ore B. Therefore, the flotation conditions were changed, and the optimum conditions developed for Ore B were applied to the blend ore. Use of Na_2S and MBS was tested at various dosages. The best results were obtained at 3000 g/t of MBS and 1000 g/t Na_2S added at the grinding stage with a mixture of Aero3418A + SIPX as collector in the copper flotation stage and with 200 g/t KAX in pyrite flotation. Table 7 shows that the copper grade and recoveries were significantly improved using the optimum flotation conditions developed for Ore B.

Table 7. Results of the open cleaner flotation test performed using a blend of Ore A:Ore B (50:50) under optimum flotation conditions (3000 g/t MBS, 1000 g/t Na_2S, 180 g/t 3418A + SIPX in copper flotation and 200 g/t KAX in pyrite flotation.

Stream	Mass Pull, %	Grade, %		Recovery, %	
		Cu	S	Cu	S
Cu Rougher Concentrate	12.88	11.04	37.67	72.28	12.24
Cu Concentrate	2.11	32.59	28.38	34.94	1.51
Pyrite Rougher Concentrate	73.63	0.71	45.84	26.56	85.16
Pyrite Concentrate	59.82	0.65	48.51	19.76	73.22
Tail	13.49	0.17	7.63	1.16	2.60
Feed				100.00	100.00

Effects of alternative conditions on flotation performance of the two ore types and their mixture were evaluated based on the results of open cleaner flotation tests. In these tests, the influence of recirculating cleaner flotation tailing was not taken into consideration. Therefore, simulation studies were performed using JKSimFloat software to estimate metallurgical performance of closed-circuit operation. In the simulation studies, first mineral assays and stage recoveries were determined by mass balance of the open cleaner flotation tests. Stage recoveries of the minerals were assumed constant in each flotation stage during simulation.

Table 8 shows the results of the simulation studies for Ore A, Ore B and their mixture under optimum flotation conditions. A copper concentrate could be produced from Ore A with approximately 28% Cu grade at 76% recovery. Use of Na_2S and MBS improved flotation response of Ore B to some extent, and a copper concentrate was produced assaying 21.62% Cu at 52.36% recovery. The copper flotation performance of the mixed ore samples was just between the two ores, as expected.

Table 8. Flotation performance of Ore A, Ore B and a mixture of the two ores under optimum flotation conditions applied on each ore type.

	Copper Concentrate			Pyrite Concentrate		
	Mass Pull (%)	Cu %	Cu Recovery (%)	Mass Pull (%)	S %	S Recovery (%)
Ore A	7.42	27.74	76.42	92.58	50.79	95.05
Ore B	2.93	21.62	52.36	63.98	44.26	92.78
Mix	4.02	31.49	65.46	77.93	49.17	93.06

Pyrite flotation was conducted following the copper flotation section. Table 8 shows that saleable grade pyrite concentrate could be produced from the three samples at high recoveries.

Surface cleaning by Na_2S and depression of pyrite by MBS mitigated the negative effect of framboidal, spongy, altered pyrite from Ore B. Following copper flotation, a high-grade pyrite concentrate could be produced from all ore samples.

4. Conclusions

Two different ore types, Ore A and Ore B from the same ore deposit, were used to investigate effects of pyrite mineralogy on the performance of the copper and pyrite flotation stages.

Mineralogical characterization showed that Ore A did not contain framboidal and altered pyrite. High grade copper concentrates could be produced at acceptable recoveries at the base conditions applied in the flotation plant.

Ore B contained framboidal and altered pyrite/marcasite and did not respond to the base flotation conditions. This was attributed to the framboidal and spongy, inclusion-rich, altered pyrite content and relatively high surface oxidation of altered marcasite particles.

Na_2S and NaHS were used for surface cleaning purpose of the sulfide minerals and control of the pulp redox potential. Both reagents improved the copper flotation performance. The copper recovery increased from about 5% up to 52%.

Blending Ore B with Ore A reduced the overall flotation performance under the base conditions. However, the flotation could be restored after sulphidization, i.e., using the optimum conditions developed for the problematic Ore B. Flotation performances of the new mix samples with different mass contents of Ore A and B can be investigated in future work.

The results of this study showed clearly the importance of process mineralogy for identification of the problematic components for flotation. Framboidal/altered pyrite particles were the main source of the problem in this case. An alternative flotation chemistry was developed based on sulphidization and the use of selective copper collectors and improved the flotation performance successfully.

Author Contributions: Conceptualization, İ.B.C. and Z.E.; methodology, İ.B.C., S.Ö. and Z.E.; validation, İ.B.C., S.Ö. and Z.E.; formal analysis, S.Ö.; investigation, İ.B.C. and Z.E.; resources, Z.E.; data curation, S.Ö. and İ.B.C.; writing—original draft preparation, İ.B.C.; writing—review and editing, Z.E. and S.Ö.; visualization, İ.B.C. and Z.E.; supervision, Z.E.; project administration, Z.E.; funding acquisition, Z.E. All authors have read and agreed to the published version of the manuscript.

Funding: This research received no external funding.

Acknowledgments: The authors would like to acknowledge the staff of ETI Copper Plant for providing the samples and their technical support.

Conflicts of Interest: The authors declare no conflict of interest.

References

1. Chandra, A.P.; Gerson, A.R. The mechanisms of pyrite oxidation and leaching: A fundamental perspective. *Surf. Sci. Rep.* **2010**, *65*, 293–315. [CrossRef]
2. Xian, Y.; Wen, S.; Chen, X.; Deng, J.; Liu, J. Effect of lattice defects on the electronic structures and floatability of pyrites. *Int. J. Miner. Metall. Mater.* **2012**, *19*, 1069. [CrossRef]
3. Anthony, J.W.; Bideaux, R.A.; Bladh, K.W.; Nichols, M.C. Pyrite. In *Handbook of Mineralogy*; Elements, Sulfides, Sulfosalts; Mineralogical Society of America: Chantilly, VA, USA, 1990; Volume I, ISBN 978-0962209734.
4. Bayraktar, İ.; Can, N.M.; Ekmekçi, Z. Effect of genesis on wetting behaviour of pyrite surfaces. In *Innovations in Mineral and Coal Processing*; Balkema: Rotterdam, The Netherlands, 1998; pp. 119–124.
5. Barker, G.J.; Gerson, A.R.; Menuge, J.F. The impact of iron sulfide on lead recovery at the giant Navan Zn–Pb orebody, Ireland. *Int. J. Miner. Process.* **2014**, *128*, 16–24. [CrossRef]
6. Zhao, C.; Huang, D.; Chen, J.; Li, Y.; Chen, Y.; Li, W. The interaction of cyanide with pyrite, marcasite and pyrrhotite. *Miner. Eng.* **2016**, *95*, 131–137. [CrossRef]

7. Wiersma, C.L.; Rimstidt, J.D. Rates of reaction of pyrite and marcasite with ferric iron at pH 2. *Geochim. Cosmochim. Acta* **1984**, *48*, 85–92. [CrossRef]
8. Butler, I.B.; Rickard, D. Framboidal pyrite formation via the oxidation of iron (II) monosulfide by hydrogen sulphide. *Geochim. Cosmochim. Acta* **2000**, *64*, 2665–2672. [CrossRef]
9. Bulatovic, S.M. *Handbook of Flotation Reagents, Chemistry, Theory and Practice, Flotation of Sulphide Ores*; Elsevier Science & Technology Books: Amsterdam, The Netherlands, 2007; p. 443.
10. Wilkin, R.T.; Barnes, H.L. Formation processes of framboidal pyrite. *Geochim. Cosmochim. Acta* **1997**, *61*, 323–339. [CrossRef]
11. Dana, J.D.; Hurlbut, C.S.; Klein, C. *Manual of Mineralogy (After James D. Dana)*; Wiley: New York, NY, USA, 1977.
12. Uhlig, I.; Szargan, R.; Nesbitt, H.W.; Laajalehto, K. Surface states and reactivity of pyrite and marcasite. *Appl. Surf. Sci.* **2001**, *179*, 222–229. [CrossRef]
13. Pugh, C.E.; Hossner, L.R.; Dixon, J.B. Oxidation rate of iron sulfides as affected by surface area, morphology, oxygen concentration, and autotrophic bacteria. *Soil Sci.* **1984**, *137*, 309–314. [CrossRef]
14. Simmons, G.L. *Flotation of Auriferous Pyrite Using Santa Fe Pacific Gold's N_2Tec Flotation Process*; SME Annual Meeting: Denver, CO, USA, 1997; Preprint 97-27.
15. Miller, J.D.; Du Plessis, R.; Kotylar, D.G.; Zhu, X.; Simmons, G.L. The low-potential hydrophobic state of pyrite in amyl xanthate flotation with nitrogen. *Int. J. Miner. Process.* **2002**, *67*, 1–15. [CrossRef]
16. Chandra, A.P.; Gerson, A.R. A review of the fundamental studies of the copper activation mechanisms for selective flotation of the sulfide minerals, sphalerite and pyrite. *Adv. Colloid Interface Sci.* **2009**, *145*, 97–110. [CrossRef] [PubMed]
17. Peng, Y.; Grano, S.; Fornasiero, D.; Ralston, J. Control of grinding conditions in the flotation of galena and its separation from pyrite. *Int. J. Miner. Process.* **2003**, *70*, 67–82. [CrossRef]
18. Sui, C.C.; Brienne, S.H.R.; Xu, Z.; Finch, J.A. Xanthate adsorption on Pb contaminated pyrite. *Int. J. Miner. Process.* **1997**, *49*, 207–221. [CrossRef]
19. Moslemi, H.; Gharabaghi, M. A review on electrochemical behavior of pyrite in the froth flotation process. *J. Ind. Eng. Chem.* **2017**, *47*, 1–18. [CrossRef]
20. Owusu, C.; Abreu, S.B.; Skinner, W.; Addai-Mensah, J.; Zanin, M. The influence of pyrite content on the flotation of chalcopyrite/pyrite mixtures. *Miner. Eng.* **2014**, *55*, 87–95. [CrossRef]
21. Petruk, W. Applied mineralogy to related gold. In *Applied Mineralogy in the Mining Industry*; Elsevier: Amsterdam, The Netherlands, 2000; Chapter 6; p. 287.
22. Boon, M.; Brasser, H.J.; Hansford, G.S.; Heijnen, J.J. Comparison of the oxidation kinetics of different pyrites in the presence of Thiobacillus ferrooxidans or Leptospirillum ferrooxidans. *Hydrometallurgy* **1999**, *53*, 57–72. [CrossRef]
23. Mu, Y.; Peng, Y.; Lauten, R.A. The depression of pyrite in selective flotation by different reagent systems—A Literature review. *Miner. Eng.* **2016**, *96–97*, 143–156. [CrossRef]
24. Bulut, G.; Ceylan, A.; Soylu, B.; Goktepe, F. Role of starch and metabisulphite on pure pyrite and pyritic copper ore flotation. *Physicochem. Probl. Miner. Process.* **2011**, *48*, 39–48.

Article

Mineralogical Appraisal of Bauxite Overburdens from Brazil

Leonardo Boiadeiro Ayres Negrão, Herbert Pöllmann * and Tiago Kalil Cortinhas Alves

Department of Mineralogy, Institute of Geosciences and Geography, Martin-Luther Halle-Wittenberg University, 06108 Halle, Germany; boiadeiro.negrao@gmail.com (L.B.A.N.); tiago.cortinhas-alves@geo.uni-halle.de (T.K.C.A.)
* Correspondence: herbert.poellmann@geo.uni-halle.de

Abstract: Mineralogical appraisal is an important tool for both mining and industrial processes. X-ray powder diffraction analysis (XRPD) can deliver fast and reliable mineralogical quantification results to aid industrial processes and improve ore recoveries. Furthermore, X-ray fluorescence (XRF) chemical data, thermal analysis (TA), and Fourier-transformed infrared spectroscopy (FTIR) can be used to validate and refine XRPD results. Mineralogical assessment of non-traditional ores, such as mining wastes, is also an important step to consider them for near-future industries. In the Brazilian Amazon, alumina-rich clays cover the largest and most important bauxitic deposits of the region and have been considered as a possible raw material for the local cement and ceramic industry. In this work, a mineralogical evaluation of these clays (Belterra Clays) is performed using XRPD, XRF, TA, and FTIR. XRPD-Rietveld quantification confirmed that kaolinite is the main phase of the clay overburden, followed by variable contents of gibbsite and goethite and minor quantities of hematite, anatase, and quartz. The chemistry derived from Rietveld, based on stoichiometric phase compositions, presents a good correlation with the XRF data and is also supported by the TA and FTIR data. The initially assumed homogeneous composition of Belterra Clay is revealed to be variable by the present mineralogical study.

Keywords: bauxite overburden; Belterra Clay; mineralogical quantification; Rietveld analysis

Citation: Negrão, L.B.A.; Pöllmann, H.; Cortinhas Alves, T.K. Mineralogical Appraisal of Bauxite Overburdens from Brazil. *Minerals* **2021**, *11*, 677. https://doi.org/10.3390/min11070677

Academic Editor: Kristian Ufer

Received: 17 May 2021
Accepted: 23 June 2021
Published: 24 June 2021

Publisher's Note: MDPI stays neutral with regard to jurisdictional claims in published maps and institutional affiliations.

1. Introduction

Chemical and phase assemblage characterizations are essential steps during industrial and mining processes to aid and improve ore processing. Traditional wet chemistry has been a reference method used for both ore and industrial products. However, these analyses are very expensive, known to consume large amounts of chemical reagents (including strong acids), and relatively slow. X-ray fluorescence (XRF) is currently one of the most used methods to determine the chemical composition of many ore types [1]. Nevertheless, XRF is not always the best option to obtain a comprehensive evaluation, as mineralogical information is necessary to understand the chemical distribution within the mineral phases and thus to evaluate ore deportment.

Mineralogical characterization is usually performed by X-ray powder diffraction (XRPD) analysis, which, when coupled with the Rietveld method, [2] enables the delivery of fast and reliable mineral phase quantifications. Online XRPD and XRF analyses are already standardized in the cement industry [3–5]. In the mining industry, the mineralogical evaluation of a variety of ores is commonly carried out by XRPD analysis, including Fe-[6,7], Ni (Co)- [8,9], Au- [10], Cu- [11,12], and Al-ore (bauxites) [13–16]. Other techniques are used to validate or even complement XRPD mineralogical analysis, such as XRF, Scan Electron Microscopy (SEM), Thermal Analysis (TA), and Fourier-transformed infrared spectroscopy (FTIR). Moreover, in many cases, assessment of ores and industrial products is performed alongside the characterization of mining residues, i.e., coal-related clays [17] and Al-refining tailings [18]. Some of these residues have been referred to as alternative raw materials, and their characterization is one of the first steps to evaluate their usage in the industry.

Bauxites, the main aluminum ore, have their largest reserves in Central and West Africa, Australia, Vietnam, and Brazil [19]. Open-pit mining of bauxite implies the removal of the bauxite overburden material, which might be shallow, as in Australia [20] and some bauxite deposits of West Africa [21,22], or relatively thick (>2 m), as in Brazil [23–25]. The bauxite overburden material in Brazil includes some topsoil and lateritic crusts, as well as thick clays that cover most bauxite deposits of the Brazilian Amazon region. These thick overburdens are usually considered a drawback during mining, as their removal increases the cost-effectiveness of the whole mining process. The bauxite overburdens in the Brazilian Amazon region are known as Belterra Clays (BTC, Figure 1) [23–26].

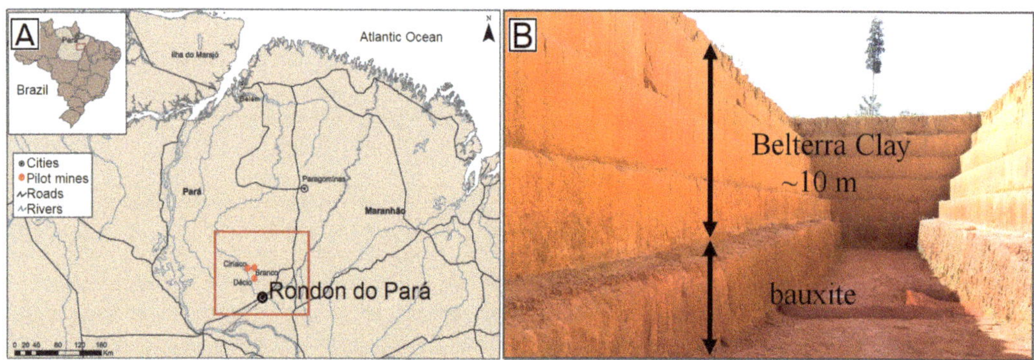

Figure 1. (**A**) Samples location; (**B**) Bauxite mining trench in Rondon do Pará (Brazil) showing the thick Belterra Clay.

Belterra Clays are dominated by kaolinite and locally have high contents of gibbsite, with minor hematite, goethite, and anatase. Their high alumina contents, associated with broad availability and low exploitation costs, have turned attention to BTC as a possible near-future raw material for the local industry. Industrial applications of these clays for the production of red ceramics [27,28] and eco-friendly cements have recently been gaining attention [29]. In this work, we performed a detailed mineralogical evaluation of Belterra Clays from different locations within the bauxite deposits of Rondon do Pará in the Brazilian Amazon region. XRPD analysis was used as the basis of the mineralogical monitoring and is discussed in light of XRF, TA, and FTIR data. These data provide an important database for the potential use of these overburdens for the production of CO_2-reduced cementitious materials.

2. Materials and Methods

For this study, we used eleven samples of Belterra Clay from three different pilot mines performed to test the bauxite ore in Rondon do Pará (Eastern Brazilian Amazon). The samples were collected in selected representative parts of the exposed clayey overburdens, with at least one sample from the top, one from the middle, and one from the base of the sequences in each pilot mine.

The samples are named according to their location and depth (in meters) within the BTC packet, and consist of three samples from the Branco bauxite pilot-mine (BRA0.5m, BRA5.0m, and BRA10m), three from the Décio pilot-mine (DEC0.8m, DEC7.2m, and DEC10m), and five samples from the Ciríaco pilot-mine (CIR1.0m, CIR5.0m, CIR7.5m, CIR10m, and CIR12m). All samples were carefully grinded to powders with clay size in an agate mortar until all grains were consistently fine for analysis. The analysis was performed using the following techniques.

XRPD analysis was performed using a Panalytical X'Pert Pro MPD X-ray diffractometer (Panalytical, Halle, Germany) equipped with a Cu anode, operated with 45 kV and 40 mA, set with a Ni Kβ-filter, and linear X'Celerator RTMS detector in the θ-θ Bragg-Brentano-Geometry. The samples were mounted by back-loading in 16 mm diameter sample holders and measured under 5–70° 2θ angular range, with a 0.013° step size, and at 38 s per step. Despite the use of a Cu source and the samples having relatively high iron contents, XRPD diffractograms had a good resolution with low background intensity and, therefore, were considered suitable for the Rietveld analysis. The data were evaluated with the software Highscore 4.5 Plus (Panalytical, Halle, Germany) from Panalytical. Principal component analysis (PCA) found large systematic variances (eigenvalues) in the set of observed samples. The method used a correlation matrix to display the most important eigenvalues (or Principal Component, PC) in a 3D plot. PCA was used to cluster similar diffractograms by similar profiles and peaks, using a Euclidian average linkage and a cut-off of 9.5. The closest sample of each sphere centroid (centroid method) was chosen as the representative sample of each group. The PDF-4 mineralogical database was used for the mineralogical characterization.

Rietveld refinement was used for the mineralogical quantification of the selected samples, which were spiked with 10% of highly crystalline fluorite to quantify the amorphous content using the internal standard method. The crystal structures (Table 1) used in the refinement were obtained from the ICSD (Inorganic Crystal Structure Database, from FIZ Karlsruhe, Germany). The accuracy of the Rietveld refinements was verified according to [30] by mixing the samples BRA13m with hematite and the sample BRA0.5m with gibbsite at 99:1, 95:5, and 90:10 ratios.

Table 1. Phases and their respective ICSD codes used in the Rietveld refinements.

Phase	Chemical Formula	Space Group	ICSD Code	Reference
Anatase	TiO_2	$I4_1/amd$	92363	[31]
Fluorite	CaF_2	Fm-3m	60368	[32]
Gibbsite	$Al(OH)_3$	$P2_1$	6162	[33]
Goethite	$(Fe,Al)O(OH)$	Pnma	109411	[34]
Hematite	Fe_2O_3	R-3c	82137	[35]
Kaolinite	$Al_2Si_2O_5(OH)_4$	P1	63192	[36]
Quartz	SiO_2	P3121	16331	[37]

The refinements were performed with the software Highscore 4.5 Plus after preparing, measuring, and refining each sample in triplicate. The scale factor, the zero shift, and the unit cells were systematically refined. The Pseudo-Voigt profile function and W profile parameters were used to better refine the peak shapes.

For comparison purposes, the sample CIR14m was also refined with the software Profex BGMN [38], with a model for disordered kaolinite downloaded from the BGMN database (Kaolinitedis.str, in http://www.bgmn.de/download-structures.html, accessed on 4 February 2021). The stacking faults in the structure of kaolinite (as b/3 shifts and layers rotations within the structure) is refined with the sub-phase approach of the software, where the disordered kaolinite is described by two sub-phases that share the same lattice parameters but have individual broadening models to describe the hkl-dependent broadening in a phenomenological way [39].

FTIR spectroscopy was also used for the mineral characterization. Approximately 1 mg of dried sample was ground with 160 mg of KBr and pressed into a disk shape. The transmission IR spectra were recorded between 370 and 4000 cm^{-1} at room temperature, with a 4 cm^{-1} resolution, using a Bruker Tensor II TGA-IR spectrometer (Bruker, Billerica, MA, USA).

XRF was used to measure the chemical composition of the selected BTC samples. Four fused pearls were prepared by mixing 1 g of sample and 8 g of fondant ($Li_2B_4O_7$) and measured in a Bruker SRS 3000 (Bruker, Billerica, MA, USA) sequential spectrometer. The loss on ignition (LOI) of the samples was determined after calcination at 1000 °C for one hour (Supplementary Materials, Table S1).

Thermal characterization of the selected samples was performed by thermogravimetric (TGA) and differential scanning calorimetric (DSC) analysis, using a Netzsch STA 449 F3

Jupiter® (Netzsch, Selb, Germany). Approximately 20 mg of sample was heated from 35 to 1000 °C in alumina crucibles under Ar-atmosphere, using a heating rate of 10 K/min.

3. Results

PCA analysis of the XRPD patterns enabled classifying them into three groups using the centroid method (Figure 2). Group 1 (green sphere) includes all BTC samples from Branco (BRA0.5m, BRA5.0m, and BRA13m) and all from Décio (DEC0.8m, DEC7.2m and DEC10m). Group 2 (blue sphere) is formed by the samples CIR0.5m, CIR5.0m, and CIR7.5m of Círiaco, whereas group 3 (gray sphere) includes the samples CIR10m and CIR14m from Círiaco.

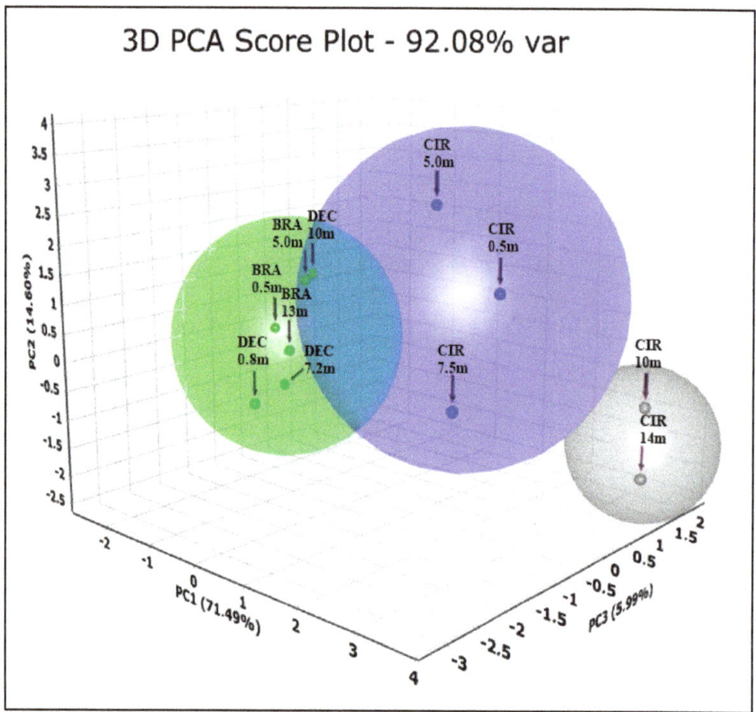

Figure 2. Principal component analysis showing the grouping of the diffractograms. Group 1: green sphere (BRA0.5m, BRA5.0m, BRA13m, DEC0.8m, DEC7.2m, and DEC10m); Group 2: blue sphere (CIR0.5m, CIR5.0m, and CIR7.5m); Group 4: gray sphere (CIR10m and CIR14m).

The samples BRA0.5m and BRA13m were chosen as representative of group 1, whereas CIR0.5m was chosen for group 2, and CIR14m was chosen for group 3. These samples were quantified with the Rietveld method, and their mineralogical assemblage consists of kaolinite, gibbsite, goethite, hematite, anatase, and quartz, which vary only in contents (Figure 3). The only exceptions are quartz, which occasionally occurs in some samples, and hematite, which is present only in the samples closest to the surface (BRA0.5m, DEC0.8m, and CIR1.0m).

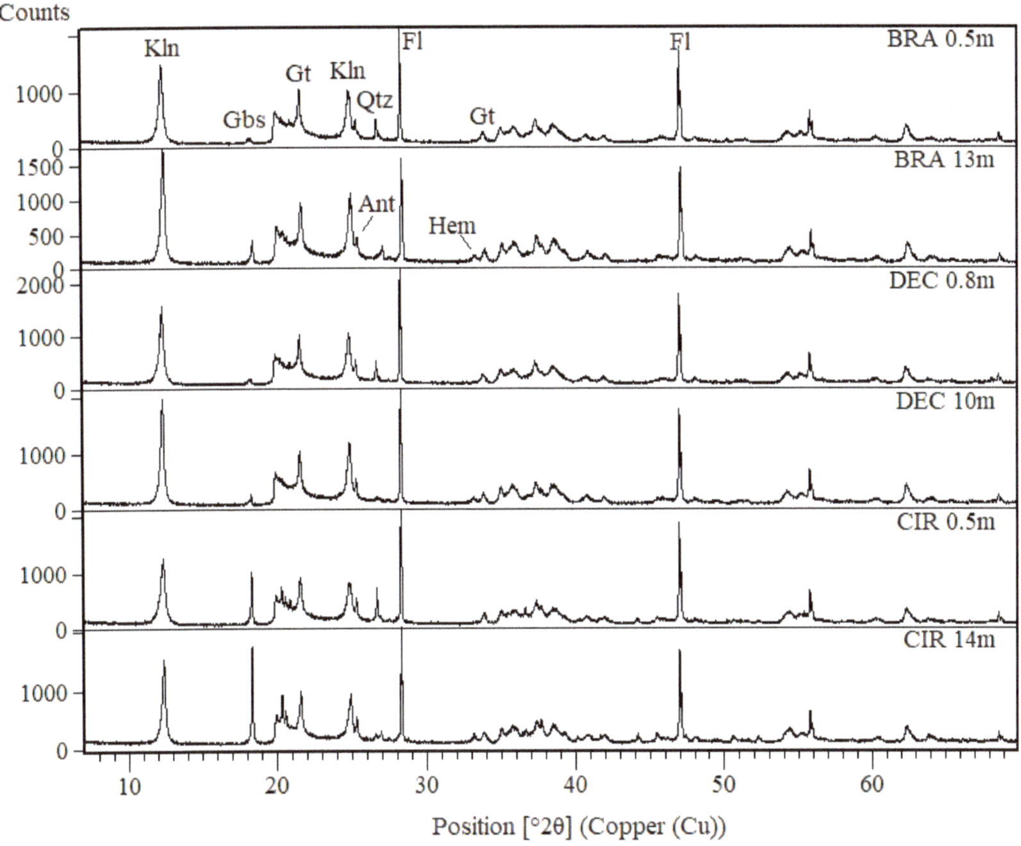

Figure 3. X-ray diffractograms of selected BTC samples spiked with fluorite (Fl). Kln: kaolinite; Gbs: gibbsite; Gt: goethite; Hem: hematite; Ant: anatase; and Qtz: quartz.

The Rietveld refinements confirm that kaolinite is the main mineral phase of the material (Figure 4, Supplementary Materials, Table S2), reaching up to 76.5% (BRA0.5m) and a minimum of 60.4% (CIR14m). Goethite varies slightly among the samples, from 14.8% to 15.9%. Gibbsite shows a stronger variation, from 3.2% to 19.2%, with higher contents in the Ciríaco mine samples. Hematite is absent in the samples close to the surface (0.5m) and represents close to 2% of the deepest samples. Anatase also shows only minor fluctuations, from 2.1% to 2.3%. Up to 2.6% of quartz was quantified, but it occurs rather occasionally as sparse quartz grains in BTC [25].

Even though the Rietveld results do not show a strong variation in the quantitative mineralogy, the Rietveld-calculated patterns presented visible misfittings in the regions 19.6–26 2θ (4.5–3.4 Å) and 34.5–37 2θ (2.6–2.4 Å) (Figure 5). These are essentially related to the refinement of kaolinite, which presents anomalous diffraction bands due to its low-ordered configuration. The disorder in kaolinite structure is mainly caused by faults in layer stacking [40]. In order to overcome this issue when refining bauxite samples rich in low-ordered kaolinites, Paz et al. [13] successfully used a calibrated hkl phase to refine this mineral in bauxites. Nevertheless, depending on the degree of disorder, an hkl phase might not accurately fit the XRPD pattern of significantly disordered kaolinites.

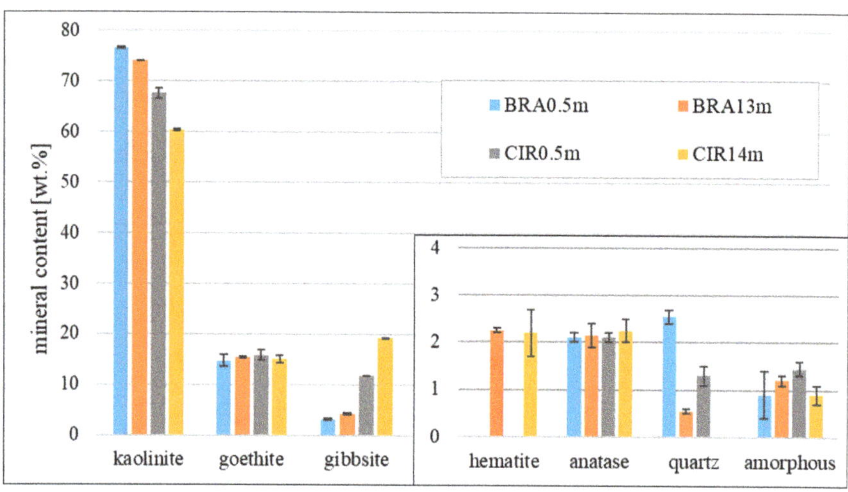

Figure 4. Mineralogical quantification of selected samples by XRPD-Rietveld. The results are the triplicate average shown with the standard deviation bars.

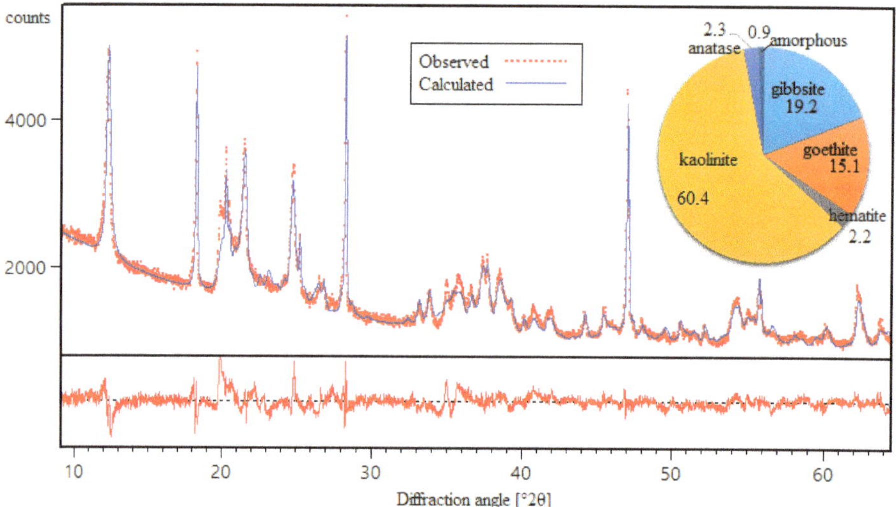

Figure 5. Refined X-ray pattern of sample CIR14m and its difference plot using Highscore Plus. Goodness of fit: 1.6 and R—weighted profile: 5.5.

An improved Rietveld-calculated fitting was achieved when using a model for disordered kaolinite (Figure 6). When compared to the previous results (Figure 5), the quantified mineral contents using the model diverge little for most of the phases (differences are close to 1%). However, kaolinite and the amorphous show a greater discrepancy. The kaolinite content is 4.6% lower when using the model for disordered kaolinite, and the amorphous is proportionally 3.7% higher. These differences are most likely related to the amorphous quantified using the internal standard method, where the over-quantified content of fluorite (internal standard) from its known added amount (10%) is recalculated as amorphous. Referencing the content, for 10.46% of fluorite quantified by the Rietveld method, the indirect calculated amorphous will be 4.6% (Figure 5), whereas it would be 0.9% (Figure 6) for a quantification of 10.09% of fluorite in the sample. Therefore, even minor differences in the first decimal case will strongly influence the amorphous quantifications when using the internal standard method. Nevertheless, 62% of kaolinite was stoichiometrically calculated after XRF (considering that all SiO_2 is in this phase), which is similar to the amount obtained in the refinement without the model (60.4%, Figure 5) and indicates that the amorphous composition in both cases (whether 0.9% or 4.6%) should have a composition close to that of kaolinite.

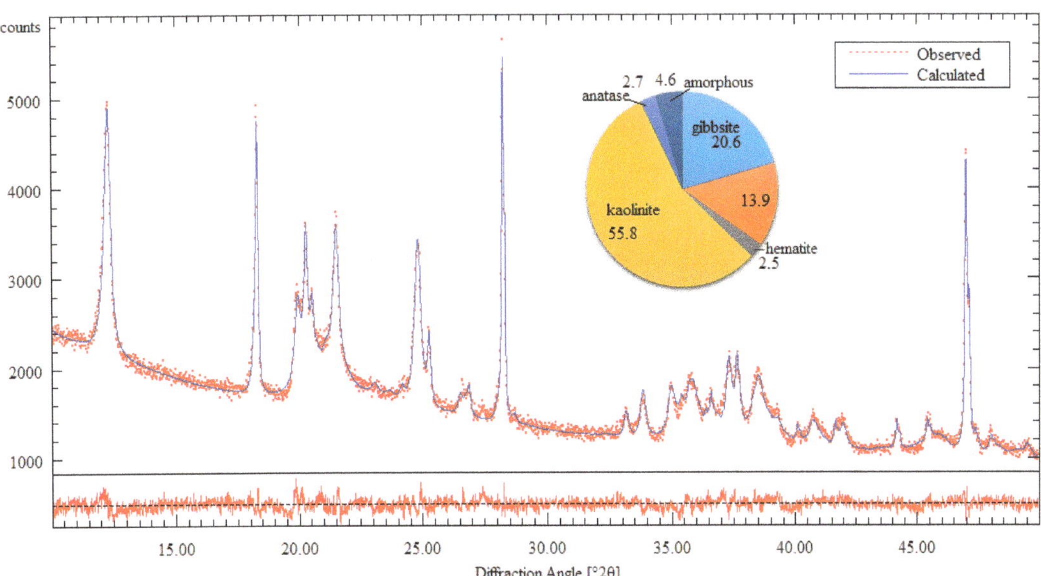

Figure 6. Refined X-ray pattern of sample CIR14m, refined in PROFEX-BGMN using a model of disordered kaolinite. Good of fitness: 1.3 and R—weighted profile: 3.2.

Besides the improvement in the Rietveld fit using the model of disordered kaolinite, the refinement was more unstable and time-consuming due to the increased interactions among the Rietveld parameters. Another attempt to better understand the contribution of defects in the structure of kaolinite was presented by Leonardi and Bish [41], who simulated the XRPD patterns of kaolinites with different layer-stacking defects and used them to analyze the patterns of natural samples. They observed that the 00*l* reflections are solely affected by the quantity of layers in the crystals, whereas the hkl reflections are also related to the amount and the nature of stacking defects, for which the stacking distance, the lateral indentation, the structure misorientation, and the structure shift were accounted. However, while the work helps to understand the nature of stacking faults in kaolinites and their contributions to their XRPD profiles, such complex profile simulations still require the use of supercomputers.

The hematite and gibbsite mixture additions proved the accuracy of the refinements (Figure 7). For both mixtures, measured in triplicate, R^2 is higher than 0.99. The resulting linear function permitted knowing the calibrated amounts of gibbsite and hematite in the samples (when *x* = *0*). These are 3.15% of gibbsite in the sample BRA0.5m and 1.99% of hematite in the sample BRA13m, values very close to the 3.1% and 2.2% quantified in the original samples, respectively. A wider standard deviation is observed for the addition of 5% of gibbsite, as a result of the preferred orientation of (00*l*) planes of gibbsite crystallites, which could not be completely solved using the March–Dollase function in the Rietveld refinement.

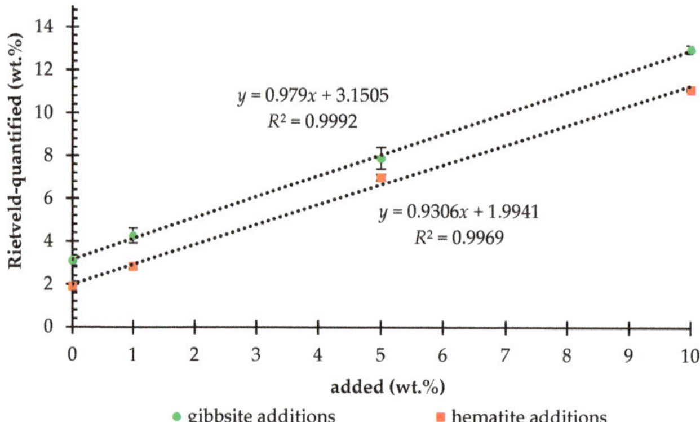

Figure 7. Mixture additions of gibbsite for the sample BRA0.5m and hematite for the sample BRA13m.

Fourier-transformed infrared spectra of the Belterra Clay samples BRA13m and CIR14m are very similar (Figure 8), showing the same bands with different intensities. Modeled and well-ordered kaolinites have well-defined OH stretching bands at 3697 (strong), 3669 (weak), 3652 (weak), and 3620 cm^{-1} [42,43]. Except for the weak band at 3669 cm^{-1}, these other bands are present in the FTIR spectra of BTC and confirm the presence of kaolinite. The disappearance of the 3669 cm^{-1} band was observed in low-ordered kaolinites by Brindley et al. [44]. These authors also reported a stronger 3652 cm^{-1} band in low-ordered kaolinites, which is a characteristically strong band in dickites and related to a disorder in the structure of kaolinite caused by the displacement of the octahedral sites' vacancies.

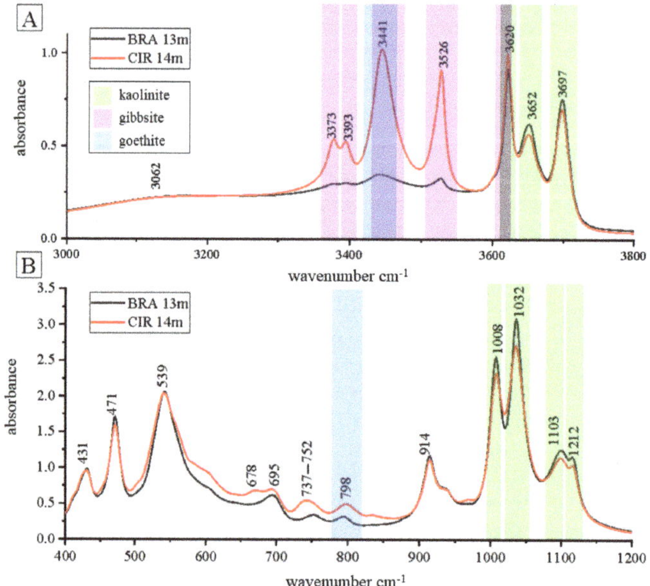

Figure 8. FT-IR spectra of BTC samples BRA13m and CIR14m: (**A**) OH-stretching region and (**B**) middle IR range. The IR bands characterized for each mineral are marked by color for a quicker interpretation. Overlapping colors means the bands are attributed to more than one mineral. Non-marked bands are related to the displacement contribution of Si, Al, and H.

The band at 3620 cm^{-1} is also an intense and characteristic band of gibbsite OH stretching, which presents its further common bands at 3526, 3441 (wider), 3393, and 3373 cm^{-1} [45]. The bands attributed to gibbsite are much more intense for CIR14m due to its higher gibbsite content. The wide band with a maximum of 3441 cm^{-1} also coincides with a non-stoichiometric OH stretching of goethite [46,47]. This mineral presents the less obvious band at ~3062 cm^{-1}, attributed to stoichiometric OH.

In the mid-IR range (Figure 8B) of the spectra, the bands at 1112, 1103, 1032, and 1008 cm^{-1} are characteristic of the Si-O stretching modes of kaolinite, whereas the band at 798 is due to the OH bend of goethite. The further bands are mainly related to the contribution of the displacement of Si, Al, and H.

The thermal analysis (Figure 9) shows four main mass-losses resulting from the heating of the BTC samples. They are well-evidenced by both DTG and DSC peaks with maximums at 50 °C, 270 °C, 375 °C, and 535 °C. The first is related to the loss of residual water absorbed by the BTC grains, whereas the following ones represent gibbsite, goethite, and kaolinite. Gibbsite dehydroxylates from 240 to approximately 380 °C, forming ρ-Al$_2$O$_3$ [48], with a consequent mass-loss of 5.49% for the CIR14m sample and 1.8% for BRA13m. The mass-loss from 340 to 410 °C corresponds to the dehydroxylation of goethite [46,47], releasing approximately 1.85% of water. The last mass-loss, 59.5% in the CIR14m and 71.9% in BRA13m, is attributed to kaolinite's decomposition to form metakaolinite [49], which later goes to mullite at approximately 980 °C [50]. The thermal analyses not only confirm the main mineralogy of the BTC samples, but are also consistent with their relative mineralogical abundancy. Using the observed mass-losses to stoichiometrically calculate mineral contents, 59.5% of kaolinite, 15.8% of gibbsite, and 16.2% of goethite are estimated for CIR14m. For BRA13m, 71.9% of kaolinite, 3.8% of gibbsite, and 16.2% of goethite are computed after the mass-losses. The calculated results are a good approximation to those obtained by the Rietveld analysis. Minor differences might be due to low-ordered and non-stoichiometric phases, such as goethite, known to be rich in Al [15]. For instance, an Al-free goethite [FeO(OH)] has 10.14% of H$_2$O, whereas an Al-rich goethite [(Fe$_{0.66}$Al$_{0.34}$)O(OH)] has 11.73% of H$_2$O to be stoichiometrically balanced. Furthermore, non-stoichiometric hydroxyl units normally occur incorporated into the structure of goethites and were found to increase proportionally to the Al/Fe ratio in this mineral [46,47].

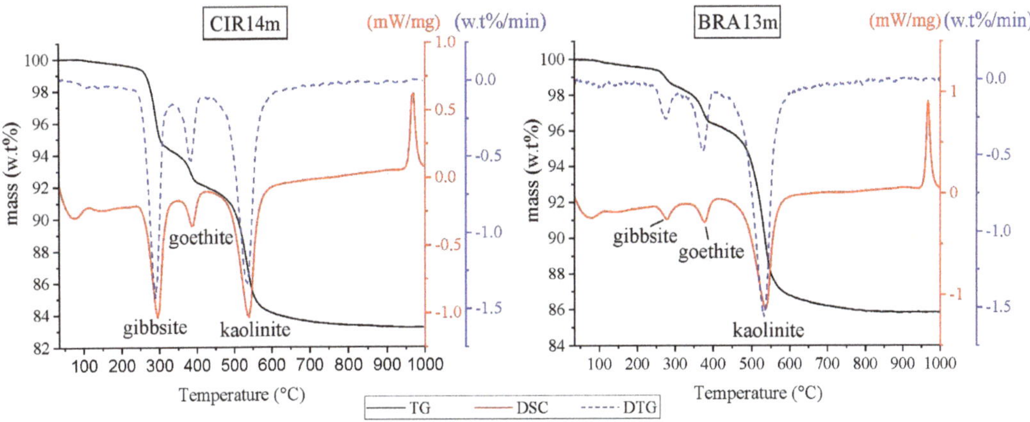

Figure 9. Thermal analysis of the BTC samples CIR14m and BRA13m. Heating rate 10 K/min. TG: termogravimetric curve; DTG: first derivate of TG and; DSC: differential scanning calorimetry.

4. Discussions

The influence of the different minerals on components of the PCA analysis is easily seen in the mineralogical quantification of the representative samples of each PCA group (Figure 4). The major difference concerns their kaolinite vs. gibbsite contents, which are proportional to the intensity of the main XRPD peaks of these minerals. Minerals with concentrations lower than 3% (quartz, hematite, and anatase, right portion of Figure 4) have some contrasting values within the same group (i.e., BRA0.5m and BRA13m of group 1). Nevertheless, these changes in minor contents (~1%) did not affect the PCA grouping, whereas the changes for kaolinite and gibbsite contents (over 10%) controlled the PCA sorting. If one wishes to consider the changes caused by minerals in minor proportions, the defined cut-off of PCA must be reduced, and the clustering would result in PCA groups composed of two or even only one scan, therefore not reducing the amount of data.

In XRPD, PCA can also be used to build calibration curves for specific minerals by inserting XRPD patterns of pure mineral phases (for example, varying from 0% to 100%) added to series of samples. However, a kaolinite standard with a XRPD pattern similar to ours is necessary to build a calibration curve for our samples, which is challenging due to the varying crystallinity of it.

The low-ordered character of kaolinite was confirmed by the different techniques used. Poorly defined peaks, or even the absence of them, were noticed in the XRPD patterns, which were better fitted in the Rietveld refinements using a special model of low-ordered kaolinite. The OH bands of kaolinite observed in the FTIR spectra are also typical of low-ordered ones. Nevertheless, a crystallinity estimation (i.e., crystallinity index) for kaolinite from the XRPD patterns [51,52] was not possible in this work, as the ($02l$) and ($11l$) reflexes of kaolinite in Belterra Clay are partially overlapped by the reflexes of gibbsite and goethite.

The chemical composition and LOI of the samples, calculated after the Rietveld results (Table S1), are plotted against the ones measured by XRF in Figure 10. The results show a good correlation with $R^2 > 0.9$ for the most abundant components (Al_2O_3, SiO_2, Fe_2O_3, and LOI) when using the following stoichiometric compositions: for kaolinite, $Al_2Si_2O_5(OH)_4$; for gibbsite, $Al(OH)_3$; and for goethite, $(Fe_{0.66}Al_{0.34})O(OH)$, which is close to the average composition of BTC goethites in Rondon do Pará [15,25]. Instead, if Al-free goethite is considered, this correlation is much poorer for Fe_2O_3 ($R^2 = 0.8334$). The poorest correlation is notably for TiO_2 ($R^2 = 0.6297$), which shows very low content and is solely influenced by the anatase. Due to its small quantities, this phase is consequently more influenced by expected errors of the Rietveld refinement [53].

Belterra Clay has indeed a similar chemical composition among the studied samples, except for Al_2O_3, which shows a variation of 4.8% from the BTC at Círiaco mine (CIR14m) to the BTC at Branco mine (BRA0.5m). Such variable alumina contents will further influence BTC's possible applicability. For ceramic purposes [27], an increase in alumina is usually related to an increase in refractoriness [54]. When applied to the production of calcium sulphoaluminate-based cements (CSAs), BTCs richer in alumina will favor the formation of calcium sulphoaluminates, whereas those with lower alumina (and consequently higher SiO_2) will result in CSA cements richer in calcium silicates [29].

Finally, the variations in SiO_2 and Al_2O_3 are mostly controlled by kaolinite and gibbsite in Belterra Clay, as the higher the gibbsite-to-kaolinite ratio is in BTC, the more enriched in alumina it will be. Gibbsite is mostly found in bauxitic nodules that occur within the clayey fraction of Belterra Clays [15,24,25,55].

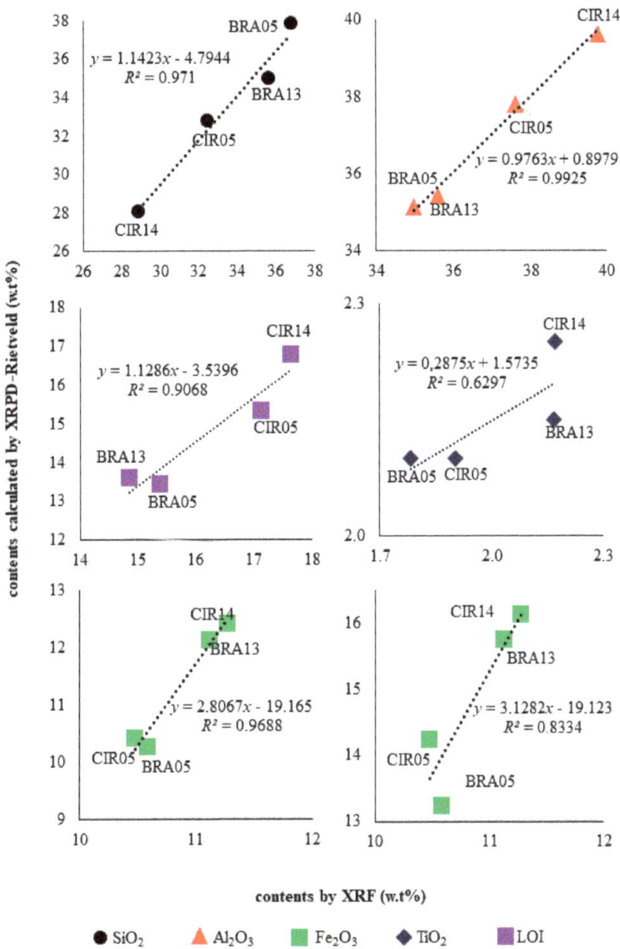

Figure 10. Comparison of the chemical composition measured by XRF and calculated after the Rietveld analysis.

5. Conclusions

The mineralogical evaluation of the clay-rich overburden of bauxites from Brazil based on XRPD analysis shows good agreement with the independently acquired results obtained by XRF chemical analysis, FTIR, and TA. When compared to traditional wet chemistry methods, the use of XRPD analysis is enhanced by the possibility of acquiring data faster, using simpler and cheaper sample preparation. Applied to BTC, the Rietveld analysis was able to deliver reliable mineralogical and chemical-derived results for the main oxides.

XRF, FTIR, and TA were complementary and crucial for an initial validation of the XRPD results and for the verification of minor phases (close to 1%). After such validations, XRPD can be applied to a variety of similar samples for rapid mineralogical control. For reliability of the results, the chemistry composition calculation after Rietveld analysis must consider the possibility of solid solutions and non-stoichiometric compositions, as was seen for Al-rich goethites in BTC.

The studied BTC samples have a quasi-homogeneous composition for Fe_2O_3 and TiO_2, controlled, respectively, by goethite and anatase. On the other hand, SiO_2 and Al_2O_3 are more variable, influenced by the abundancies of kaolinite and gibbsite. The major differences in an industrial application of BTC will be further related to its alumina contents. Whereas the alumina content of Belterra Clay is too low for Al-ore (as bauxite), it is high enough to consider it as a non-traditional raw material to be used in the cement or ceramic industries.

Supplementary Materials: The following materials are available online at https://www.mdpi.com/article/10.3390/min11070677/s1, Table S1. Chemistry (wt. %) of the studied representative Belterra Clay samples, Table S2. Mineralogical composition of the studied representative Belterra Clay samples after the Rietveld-XRPD quantifications. Rwp: R-weighted profile; GOF: goodness of fit.

Author Contributions: Conceptualization, L.B.A.N. and H.P.; methodology, L.B.A.N.; software, L.B.A.N. and T.K.C.A.; formal analysis, L.B.A.N.; investigation, L.B.A.N.; resources, L.B.A.N. and H.P.; data curation, L.B.A.N. and T.K.C.A.; writing—original draft preparation, L.B.A.N.; writing—review and editing, H.P.; supervision, H.P.; funding acquisition, L.B.A.N. All authors have read and agreed to the published version of the manuscript.

Funding: This research was funded by CAPES foundation, grant number: 88881.199654/2018-01.

Data Availability Statement: The data produced is available in this publication and in its Supplementary material. Further data can be shared after request to the authors.

Acknowledgments: The authors acknowledge Marcondes Lima da Costa and the company *Nexa Resources* for the BTC samples. The financial support (first author) of the Brazilian CAPES foundation, through the grant 88881.199654/2018-01, and the valuable contributions of the editors and anonymous reviewers are also gratefully acknowledged.

Conflicts of Interest: The authors declare no conflict of interest.

References

1. Streli, C.; Wobrauschek, P.; Kregsamer, P. X-ray Fluorescence Spectroscopy, Applications. In *Encyclopedia of Spectroscopy and Spectrometry*; Elsevier: Amsterdam, The Netherlands, 1999; pp. 2478–2487.
2. Rietveld, H.M. A profile refinement method for nuclear and magnetic structures. *J. Appl. Crystallogr.* **1969**, *2*, 65–71. [CrossRef]
3. Aranda, M.A.G.; De la Torre, A.G.; León-Reina, L. Powder-diffraction characterization of cements. In *International Tables for Crystallography*; International Union of Crystallography: Chester, UK, 2019; ISBN 9780470685754. [CrossRef]
4. Galluccio, S.; Pöllmann, H. Quantifications of Cements Composed of OPC, Calcined Clay, Pozzolanes and Limestone. In *Calcined Clays for Sustainable Concrete*; Springer: Berlin/Heidelberg, Germany, 2020.
5. Khelifi, S.; Ayari, F.; Tiss, H.; Hassan Chehimi, D. Ben X-ray fluorescence analysis of Portland cement and clinker for major and trace elements: Accuracy and precision. *J. Aust. Ceram. Soc.* **2017**, *53*, 743–749. [CrossRef]
6. König, U.; Norberg, N.; Gobbo, L. From iron ore to iron sinte-Process Control using X-ray Diffraction. In Proceedings of the 45° Ironmaking/16° Iron Ore/3° Agglomeration, Rio de Janeiro, Brazil, 17–21 August 2015; pp. 146–153.
7. De Villiers, J.P.R.; Lu, L. XRD analysis and evaluation of iron ores and sinters. In *Iron Ore: Mineralogy, Processing and Environmental Sustainability*; Elsevier: Amsterdam, The Netherlands, 2015; pp. 85–100.
8. Santoro, L.; Putzolu, F.; Mondillo, N.; Herrington, R.; Najorka, J.; Boni, M.; Dosbaba, M.; Maczurad, M.; Balassone, G. Quantitative mineralogical evaluation of Ni-Co laterite ores through XRPD-QPA- and automated SEM-based approaches: The Wingellina (Western Australia) case study. *J. Geochem. Explor.* **2021**, *223*, 106695. [CrossRef]
9. Ribeiro, P.P.M.; de Souza, L.C.M.; Neumann, R.; dos Santos, I.D.; Dutra, A.J.B. Nickel and cobalt losses from laterite ore after the sulfation-roasting-leaching processing. *J. Mater. Res. Technol.* **2020**, *9*, 12404–12415. [CrossRef]
10. Ospina-Correa, J.D.; Mejía-Restrepo, E.; Serna-Zuluaga, C.M.; Posada-Montoya, A.; Osorio-Cachaya, J.G.; Tamayo-Sepúlveda, J.A.; Calderón-Gutiérrez, J.A. Process mineralogy of refractory gold ore in thiosulfate solutions. *Hydrometallurgy* **2018**, *182*, 104–113. [CrossRef]
11. Rahfeld, A.; Kleeberg, R.; Möckel, R.; Gutzmer, J. Quantitative mineralogical analysis of European Kupferschiefer ore. *Miner. Eng.* **2018**, *115*, 21–32. [CrossRef]
12. Li, G.; Cheng, H.; Xu, C.; Lu, C.; Lu, X.; Zou, X.; Xu, Q. Mineralogical Analysis of Nickel/Copper Polymetallic Sulfide Ore by X-ray Diffraction Using Rietveld Method. In *Characterization of Minerals, Metals, and Materials 2016*; Ikhmayies, S.J., Li, B., Carpenter, J.S., Hwang, J.-Y., Monteiro, S.N., Li, J., Firrao, D., Zhang, M., Peng, Z., Escobedo-Diaz, J.P., et al., Eds.; Springer International Publishing: Cham, Switzerland, 2016; pp. 67–74.
13. Paz, S.P.A.; Angélica, R.S.; Kahn, H. Optimization of the reactive silica quantification method applied to Paragominas-type gibbsitic bauxites. *Int. J. Miner. Process.* **2017**, *162*, 48–57. [CrossRef]
14. Melo, C.C.A.; Angélica, R.S.; Paz, S.P.A. A proposal for rapid grade control of gibbsitic bauxites using multivariate statistics on XRD data. *Miner. Eng.* **2020**, *157*, 106539. [CrossRef]

15. Negrão, L.B.A.; da Costa, M.L.; Pöllmann, H.; Horn, A. An application of the Rietveld refinement method to the mineralogy of a bauxite-bearing regolith in the Lower Amazon. *Mineral. Mag.* **2018**, *82*, 413–431. [CrossRef]
16. Negrão, L.B.A.; da Costa, M.L. Mineralogy and geochemistry of a bauxite-bearing lateritic profile supporting the identification of its parent rocks in the domain of the huge Carajás iron deposits, Brazil. *J. South. Am. Earth Sci.* **2021**, *108*, 103164. [CrossRef]
17. De Ruan, C.; Ward, C.R. Quantitative X-ray powder diffraction analysis of clay minerals in Australian coals using Rietveld methods. *Appl. Clay Sci.* **2002**, *21*, 227–240. [CrossRef]
18. Santini, T.C. Application of the Rietveld refinement method for quantification of mineral concentrations in bauxite residues (alumina refining tailings). *Int. J. Miner. Process.* **2015**, *139*, 1–10. [CrossRef]
19. U.S Geological Survey. *Mineral Commodity Summaries*; U.S Geological Survey: Reston, VA, USA, 2021.
20. Horbe, A.M.C.; Anand, R.R. Bauxite on igneous rocks from Amazonia and Southwestern of Australia: Implication for weathering process. *J. Geochem. Explor.* **2011**, *111*, 1–12. [CrossRef]
21. Nyamsari, D.G.; Yalcin, M.G. Statistical analysis and source rock of the Minim-Martap plateau bauxite, Cameroon. *Arab. J. Geosci.* **2017**, *10*, 415. [CrossRef]
22. Sidibe, M.; Yalcin, M.G. Petrography, mineralogy, geochemistry and genesis of the Balaya bauxite deposits in Kindia region, Maritime Guinea, West Africa. *J. Afr. Earth Sci.* **2019**, *149*, 348–366. [CrossRef]
23. Horbe, A.M.C.; da Costa, M.L. Lateritic crusts and related soils in eastern Brazilian Amazonia. *Geoderma* **2005**, *126*, 225–239. [CrossRef]
24. Truckenbrodt, W.; Kotschoubey, B.; Schellmann, W. Composition and origin of the clay cover on North Brazilian laterites. *Geol. Rundsch.* **1991**, *80*, 591–610. [CrossRef]
25. Negrão, L.B.A.; da Costa, M.L.; Pöllmann, H. The Belterra Clay on the bauxite deposits of Rondon do Pará, Eastern Amazon. *Braz. J. Geol.* **2018**, *48*, 473–484. [CrossRef]
26. Sombroek, W.G. *Amazon Soils: A Reconnaissance of the Soils Region, the Brazilian Amazonitle*; Wageningen University: Wageningen, The Netherlands, 1966.
27. Barreto, I.A.R.; da Costa, M.L. Sintering of red ceramics from yellow Amazonian latosols incorporated with illitic and gibbsitic clay. *Appl. Clay Sci.* **2018**, *152*, 124–130. [CrossRef]
28. Barreto, I.A.R.; da Costa, M.L. Viability of Belterra clay, a widespread bauxite cover in the Amazon, as a low-cost raw material for the production of red ceramics. *Appl. Clay Sci.* **2018**, *162*, 252–260. [CrossRef]
29. Negrão, L.B.A.; Pöllmann, H.; da Costa, M.L. Production of low-CO_2 cements using abundant bauxite overburden "Belterra Clay". *Sustain. Mater. Technol.* **2021**, e00299. [CrossRef]
30. Negrão, L.B.A.; Pöllmann, H. The Phase Addition method to evaluate Rietveld mineral quantitative analysis of hydrated cements. *Bol. DO Mus. Geociênc. Amaz.* **2020**, *7*, 7. [CrossRef]
31. Weirich, T.E.; Winterer, M.; Seifried, S.; Hahn, H.; Fuess, H. Rietveld analysis of electron powder diffraction data from nanocrystalline anatase, TiO_2. *Ultramicroscopy* **2000**, *81*, 263–270. [CrossRef]
32. Batchelder, D.N.; Simmons, R.O. Lattice constants and thermal expansivities of silicon and of calcium fluoride between 6 and 322 K. *J. Chem. Phys.* **1964**, *41*, 2324–2329. [CrossRef]
33. Saalfeld, H.; Wedde, M. Refinement of the crystal structure of gibbsite, $A1(OH)_3$. *Z. Krist. Cryst. Mater.* **1974**, *139*, 129–135. [CrossRef]
34. Li, D.; O'Connor, B.H.; Low, I.-M.; van Riessen, A.; Toby, B.H. Mineralogy of Al-substituted goethites. *Powder Diffr.* **2006**, *21*, 289–299. [CrossRef]
35. Sadykov, V.A.; Isupova, L.A.; Tsybulya, S.V.; Cherepanova, S.V.; Litvak, G.S.; Burgina, E.B.; Kustova, G.N.; Kolomiichuk, V.N.; Ivanov, V.P.; Paukshtis, E.A.; et al. Effect of mechanical activation on the real structure and reactivity of iron (III) oxide with corundum-type structure. *J. Solid State Chem.* **1996**, *123*, 191–202. [CrossRef]
36. Bish, D.L.; Von Dreele, R.B. Rietveld refinement of non-hydrogen atomic positions in kaolinite. *Clays Clay Miner.* **1989**, *37*, 289–296. [CrossRef]
37. D'Amour, H.; Denner, W.; Schulz, H. Structure determination of α-quartz up to 68×10^8 Pa. *Acta Crystallogr. Sect. B Struct. Crystallogr. Cryst. Chem.* **1979**, *35*, 550–555. [CrossRef]
38. Doebelin, N.; Kleeberg, R. Profex: A graphical user interface for the Rietveld refinement program BGMN. *J. Appl. Crystallogr.* **2015**, *48*, 1573–1580. [CrossRef]
39. Bergmann, J.; Kleeberg, R. Rietveld Analysis of Disordered Layer Silicates. *Mater. Sci. Forum* **1998**, *278–281*, 300–305. [CrossRef]
40. Ufer, K.; Kleeberg, R.; Monecke, T. Quantification of stacking disordered Si-Al layer silicates by the Rietveld method: Application to exploration for high-sulphidation epithermal gold deposits. *Powder Diffr.* **2015**, *30*, S111–S118. [CrossRef]
41. Leonardi, A.; Bish, D.L. Understanding Powder X-ray Diffraction Profiles from Layered Minerals: The Case of Kaolinite Nanocrystals. *Inorg. Chem.* **2020**, *59*, 5357–5367. [CrossRef]
42. Chukanov, N.V.; Chervonnyi, A.D. IR Spectra of Minerals and Related Compounds, and Reference Samples' Data. In *Principles and Applications of Well Logging*; Springer: Berlin/Heidelberg, Germany, 2016; pp. 51–1047.
43. Balan, E.; Saitta, A.M.; Mauri, F.; Calas, G. First-principles modeling of the infrared spectrum of kaolinite. *Am. Mineral.* **2001**, *86*, 1321–1330. [CrossRef]
44. Brindley, G.W.; Kao, C.C.; Harrison, J.L.; Lipsicas, M.; Raythatha, R. Relation between structural disorder and other characteristics of kaolinites and dickites. *Clays Clay Miner.* **1986**, *34*, 239–249. [CrossRef]

45. Balan, E.; Lazzeri, M.; Morin, G.; Mauri, F. First-principles study of the OH-stretching modes of gibbsite. *Am. Mineral.* **2006**, *91*, 115–119. [CrossRef]
46. Ruan, H.D.; Frost, R.L.; Kloprogge, J.T.; Duong, L. Infrared spectroscopy of goethite dehydroxylation. II. Effect of aluminium substitution on the behaviour of hydroxyl units. *Spectrochim. Acta Part A Mol. Biomol. Spectrosc.* **2002**, *58*, 479–491. [CrossRef]
47. Schulze, D.G.; Schwertmann, U. The influence of aluminium on iron oxides: X. properties of Al-substituted goethites. *Clay Miner.* **1984**, *19*, 521–539. [CrossRef]
48. Colombo, C.; Violante, A. Effect of time and temperature on the chemical composition and crystallization of mixed iron and aluminum species. *Clays Clay Miner.* **1996**, *44*, 113–120. [CrossRef]
49. Yeskis, D.; Van Groos, A.F.K.; Guggenheim, S. The dehydroxylation of kaolinite. *Am. Mineral.* **1985**, *70*, 159–164.
50. Chen, Y.F.; Wang, M.C.; Hon, M.H. Phase transformation and growth of mullite in kaolin ceramics. *J. Eur. Ceram. Soc.* **2004**, *24*, 2389–2397. [CrossRef]
51. Aparicio, P.; Galán, E.; Ferrell, R.E. A new kaolinite order index based on XRD profile fitting. *Clay Miner.* **2006**, *41*, 811–817. [CrossRef]
52. Plançon, A.; Giese, R.F.; Snyder, R. The Hinckley index for kaolinites. *Clay Miner.* **1988**, *23*, 249–260. [CrossRef]
53. Bish, D.L.; Howard, S.A. Quantitative phase analysis using the Rietveld method. *J. Appl. Crystallogr.* **1988**, *21*, 86–91. [CrossRef]
54. Djangang, C.N.; Kamseu, E.; Ndikontar, M.K.; Nana, G.L.L.; Soro, J.; Melo, U.C.; Elimbi, A.; Blanchart, P.; Njopwouo, D. Sintering behaviour of porous ceramic kaolin-corundum composites: Phase evolution and densification. *Mater. Sci. Eng. A* **2011**, *528*, 8311–8318. [CrossRef]
55. Da Costa, M.L.; da Cruz, G.S.; de Almeida, H.D.F.; Poellmann, H. On the geology, mineralogy and geochemistry of the bauxite-bearing regolith in the lower Amazon basin: Evidence of genetic relationships. *J. Geochemical Explor.* **2014**, *146*, 58–74. [CrossRef]

Article

A Method for Quality Control of Bauxites: Case Study of Brazilian Bauxites Using PLSR on Transmission XRD Data

Caio C. A. Melo [1,2,*], Rômulo S. Angélica [2] and Simone P. A. Paz [1,2]

[1] Institute of Technology, Federal University of Pará (PRODERNA/ITEC/UFPA), 66075-110 Belém, Brazil; paz@ufpa.br
[2] Institute of Geosciences, Federal University of Pará (PPGG/IG/UFPA), 66075-110 Belém, Brazil; angelica@ufpa.br
* Correspondence: eng.caiomelo@hotmail.com

Abstract: Available Alumina ($Av\text{Al}_2\text{O}_3$) and Reactive Silica ($Rx\text{SiO}_2$), the main parameters of bauxite controlled in the beneficiation process are traditionally measured by laborious, expensive, and time-consuming wet chemistry methods. Alternative methods based on XRD analysis, capable to provide a reliable estimation of these parameters and valuable mineralogical information of the ore, are being studied. In this work, X-ray diffraction data in transmission mode was used to estimate $Av\text{Al}_2\text{O}_3$ and $Rx\text{SiO}_2$ from Brazilian bauxites using the Partial Least Square Regression (PLSR) statistical tool. The proposed method comprises a routine of sample classification according to their similarities by Principal Component Analysis (PCA) and K-means, calibration of the PLSR model for each group of samples, grouping new bauxite samples according to the generated clustering model, and subsequent estimation of the parameters $Av\text{Al}_2\text{O}_3$ and $Rx\text{SiO}_2$ using the PLSR models for these samples. The results showed good accuracy and precision of the models generated for samples of the main ore lithology. The quality and pre-processing of the XRD data required for this method are discussed. The results demonstrated that this method has the potential to be industrially applied to quality control of bauxites as a rapid and automated procedure.

Keywords: bauxite; available alumina; reactive silica; XRD; PLSR

Citation: Melo, C.C.A.; Angélica, R.S.; Paz, S.P.A. A Method for Quality Control of Bauxites: Case Study of Brazilian Bauxites Using PLSR on Transmission XRD Data. *Minerals* **2021**, *11*, 1054. https://doi.org/10.3390/min11101054

Academic Editor: Giovanni Mongelli

Received: 26 May 2021
Accepted: 8 July 2021
Published: 28 September 2021

Publisher's Note: MDPI stays neutral with regard to jurisdictional claims in published maps and institutional affiliations.

1. Introduction

Bauxite is the main aluminum ore with global resources estimated to be 55–75 billion tons. Brazil holds the 4th largest reserve and produces annually 35 million tons of bauxite, mainly to produce smelter grade alumina [1].

The main aluminum-ore mineral present in Brazilian lateritic bauxite is gibbsite (known as available alumina—$Av\text{Al}_2\text{O}_3$), and therefore, these bauxites are processed in low-temperature digestion (LTD) conditions (100–150 °C) [2,3]. In this context, among the silicon-bearing minerals, only kaolinite is leached in the Bayer process. This gangue mineral is well known in the industry as reactive silica ($Rx\text{SiO}_2$) since it rapidly and undesirably reacts with the sodium hydroxide solution releasing Na_2SiO_3 to the pregnant liquor (Equation (1)), which must be precipitated as zeolitic phases known as DSP (desilication product) even during the digestion stage (Equation (2)) [4]. These neoformed products significantly affect the costs of the process, either because they dictate the time and influence the temperature of digestion, but mainly because of the loss of caustic soda, making it, in many cases, economically unfeasible to process bauxites with $Rx\text{SiO}_2 > 5\%$ [3–5].

$$3\text{Al}_2\text{Si}_2\text{O}_5(\text{OH})_{4\ (s)} + 18\text{NaOH}_{(aq)} \rightarrow 6\text{Na}_2\text{SiO}_{3\ (aq)} + 6\text{NaAl(OH)}_{4\ (aq)} + 3\text{H}_2\text{O}_{(l)} \quad (1)$$

$$6\text{Na}_2\text{SiO}_{3\ (aq)} + 6\text{NaAl(OH)}_{4\ (aq)} + \text{Na}_2\text{X}_{(aq)}$$
$$\rightarrow \text{Na}_6(\text{Al}_6\text{Si}_6\text{O}_{24})\text{Na}_2\text{X}_{(s)} + 12\text{NaOH}_{(aq)} + 6\text{H}_2\text{O}_{(l)} \quad (2)$$

In the mineral industry, it is common to have quality control and process parameters based on chemical data instead of mineralogical. It is mainly due to the consolidation and availability of quantitative chemical analysis using wet methods, while methods for mineralogical determination are still under development. In the aluminum industry, it is no different. Quality control of ore in the mine and the Bayer process is done almost exclusively in terms of its chemical composition. Thus, samples from geological research, beneficiation, and Bayer process feedstock are analyzed to determine the content of available alumina ($AvAl_2O_3$) and reactive silica ($RxSiO_2$)—traditionally determined by wet chemistry [2,6,7].

Paz et al. [2] report that the $RxSiO_2$ content determined by such methods can be underestimated, depending on the content and the degree of crystallinity of the kaolinite in bauxite, which may change significantly over the bauxite profile. This means that the clay mineral present can be more reactive to the process, despite its concentration [2,8]. Thus, there is no guarantee that simple knowledge of the chemical composition of bauxite will allow efficient control in metallurgical processes [9]. Another downside of traditional methods is that they are time-consuming, demand manpower and space, and involve the handling of dangerous reagents [10,11].

In this context, several methods based on the mineralogical composition of the ore, obtained by X-ray diffraction (XRD) are being developed as an alternative for the process control in the bauxite and alumina industry. These methods are, in general, based on powder XRD data using Rietveld refinement [6,10–17] and multivariate statistics [18–21]. Feret [13] states that XRD has become a fundamental and irreplaceable tool in the control of raw materials of the aluminum industry. The advent of high-speed XRD detectors have enabled a fast data collection and, consequently, the development of rapid and accurate methods, as they use the whole XRD pattern, reducing the effect of preferred orientation and reflection extinction and even mitigating the inaccuracies due to amorphous content [10,11,13–15]. König et al. [10] demonstrated the mineralogical quantification of certified bauxite samples from several countries. Aylmore and Walker [14] and Nong et al. [15] also applied Rietveld-XRD for the quantification of Australian lateritic and Chinese karstic bauxites, respectively. Applications of powder XRD to quantify Brazilian bauxites from Paragominas and Juruti (northern Brazil) were also studied by Angélica et al. [6] and Negrão et al. [16], respectively. Feret and See [17] reported a bauxite analysis by XRD using synchrotron radiation to improve mineralogical quantification.

Principal Component Analysis (PCA) and Partial Least Square Regression (PLSR) are two statistical methods widely used in the chemometric field [22–24]. Viscarra Rossel et al. [25,26] demonstrated the use of PLSR from UV-Vis and infrared data to predict various soil properties (such as pH, organic carbon (OC), cation exchange capacity (CEC), etc.) and to determine the composition of mineral-organic mixtures in soils, while PCA was used to compare the synthetic mixtures with respective soils. Olatunde [27] reported excellent results using PLSR on infrared data to estimate the extractible total petroleum hydrocarbon (ETPH) in soils. The author highlights the accuracy and rapidness of this method. PLSR was also used on Energy dispersive X-ray fluorescence (EDXRF) data to predict some soil parameters (CEC, sum of exchangeable bases (SB), and base saturation percentage (BSP)) [28]. From XRD data, König et al. [29] demonstrated the utilization of PLSR for quality control of iron ore sinter as a reliable, easy and rapid method in contrast to wet chemistry.

Melo et al. [19,20] developed a methodology using PLSR on XRD data (reflection geometry), applied to estimate the bauxite quality control parameters. The authors reported that the estimation of $AvAl_2O_3$ and $RxSiO_2$ obtained were in good agreement with the reference and within the acceptable limits of precision (<1.0–1.5% and <0.5%, respectively) [30]. However, it was observed that in samples of marginal ore lithologies with higher kaolinite content and degree of crystallinity (low defects kaolinite), the method does not meet the precision limits, probably due to the preferred orientation effect from manual sample preparation. To overcome this issue, this study aimed to use XRD data in transmission mode following a methodology similar to that of Melo et al. [19] applied to

Brazilian gibbsitic bauxites. It is worth mentioning that the proposed method has several advantages over the traditional methods: rapidness, can be completely automated, no chemical reagents are required, and the ore mineralogy can also be monitored providing relevant information to the process.

2. Materials and Methods

The bauxite samples were provided by Mineração Paragominas SA (Norsk Hydro) and correspond to a drilling campaign on the Miltonia 3 plateau, Pará state, northern Brazil [31]. In this study, 105 samples were used, corresponding to four lithologies: Nodular Bauxite (BN), Nodular-Crystalized Bauxite (BNC), Crystalized Bauxite (BC), and Crystalized-Amorphous Bauxite (BCBA). Details of this lithological profile and sample preparation can be found in Silva et al. [32] and Melo et al. [19]. Figure 1 depicts a schematic representation of the Miltonia location and geological profile.

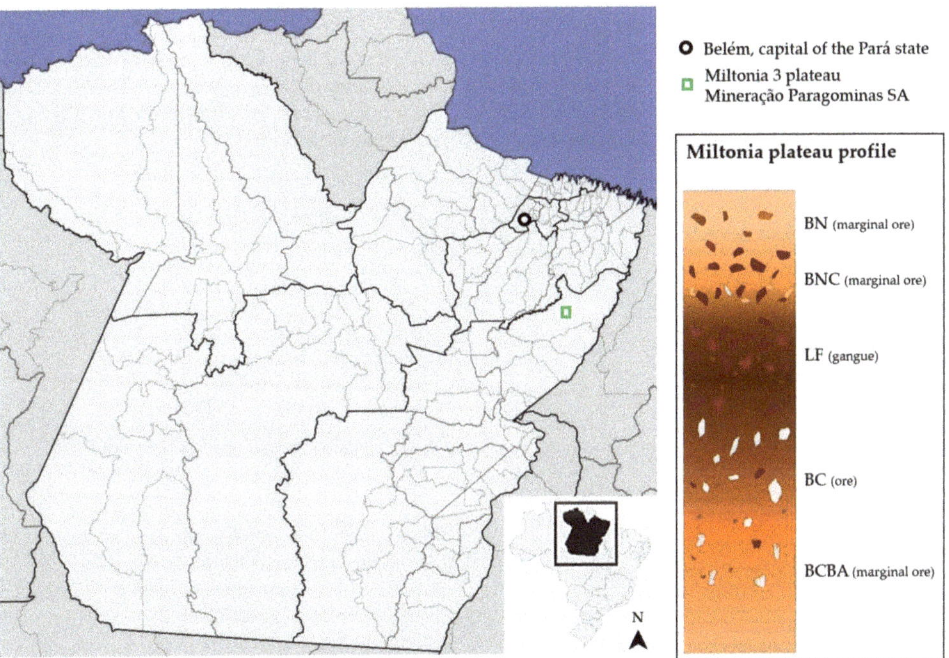

Figure 1. Location and typical geological profile of Miltonia plateau (Pará state, northern Brazil). Legend: BN: Nodular Bauxite, BNC: Nodular-Crystalized Bauxite, LF: Ferruginous Laterite, BC: Crystalized Bauxite, and BCBA: Crystalized-Amorphous Bauxite lithology.

The powder XRD data were collected using a diffractometer (Empyrean, Panalytical, Almelo, The Netherlands), Co X-ray tube ($K\alpha_1$ = 1.789 Å), Fe Kβ filter, and PIXel3D 2 × 2 area detector (linear scanning mode) with an active length of 3.3473° 2θ (255 channels). The following conditions of data collection were used: Transmission mode; 40 kV and 35 mA; soller slit of 0.04 rad; fixed divergent and anti-scattering slits of 1/8°; 0.066° 2θ step-size; 22.96 s of time/step and scanning range from 5° to 70° 2θ. The step-size was defined based on Melo et al.'s [14] optimization conditions. Diffractograms were evaluated using the software HighScore Plus 4.8 (Panalytical, Almelo, The Netherlands).

Each sample was assembled in the sample holder and analyzed in duplicate by XRD. To perform the PCA, K-means and PLSR analyses, XRD data were used as dataset. Thus, all diffractograms are organized as an $m \times n$ matrix, where m (rows) are the bauxite samples and n (columns) are the intensity count value for each $°2\theta$ step of the XRD measurement for the respective sample. Here, the complete XRD pattern is taken as dependent variables, resulting in 984 features for modeling [19].

PCA was carried out to identify possible outliers and samples with mineralogical similarity; and K-means clustering algorithm (with $k = 3$, considering Euclidean distance measure) was used to group the samples with similarities (the clusters were named as C1, C2, and C3).

The samples classified in each cluster were randomly divided into two subsets: a calibration set (containing ~70% of the samples) and a test set (~30% of the samples). The samples from the calibration set were used to build the PLSR models. This statistical algorithm is particularly suitable for handling multi-collinear data, and an interesting alternative for predicting relevant information Y (obtained from expensive, difficult, or time-consuming measurements—e.g., wet chemistry) from X data (in general, cheap, easy, or fast measurements—e.g., XRD, Fourier Transform Infrared Spectroscopy (FTIR)) [18,22,33]. Thus, in this study, the content of $AvAl_2O_3$ and $RxSiO_2$ (from wet chemistry) was predicted by using XRD data.

A "leave-one-out" cross-validation was used to find the best number of factors to include in the models and the Root Mean Square Error of Prediction (RMSEP, Equation (3)), Ratio of Prediction Deviation (RPD, Equation (4)), and Relative Error (RE, Equation (5)) were used to assess the performance of the models.

$$RMSEP = \sqrt{\frac{\sum_{i=1}^{n}\left(y_{i,\ predicted} - y_{i,\ reference}\right)^2}{n}} \tag{3}$$

$$RPD = \frac{SD_{prediction\ set}}{RMSEP} \tag{4}$$

$$RE\ (\%) = 100\sqrt{\frac{\sum_{i=1}^{n}\left(y_{i,\ predicted} - y_{i,\ reference}\right)^2}{\sum_{i=1}^{n}y_{i,\ reference}^2}} \tag{5}$$

3. Results and Discussion

3.1. XRD Data

Figure 2 shows the X-ray diffractograms of all the bauxite samples used in this study collected by transmission mode. It can be noted that the bauxites of the four lithologies have the same mineralogical composition. The main phase is gibbsite (d_{002} = 4.85 Å and d_{110} = 4.37 Å). In general, the only SiO_2 mineral identified is kaolinite (d_{001}~7.14 Å and d_{002}~3.58 Å). Quartz (d_{101} = 3.34 Å) may be present in some samples, but in minor content. Hematite (d_{104} = 2.69 Å) is observed as the main iron mineral, with intensity varying significantly among the lithologies, and Al-goethite (d_{101} ranging from 4.18 Å to 4.14 Å) is also observed, usually as a broad peak due to variations in the isomorphic substitution of Al in the structure [12,34]. Anatese (d_{101} = 3.52 Å) is also present in all samples.

Layered minerals (such as clay minerals) tend to orient themselves strongly during samples' assembling in the sample holders for XRD analysis. Thus, for those samples rich in kaolinite and/or gibbsite, it is common to observe high intensities of the basal reflections (d_{001}) in detriment of the other reflections of the XRD pattern [35]. This effect is believed to be the major source of error in quantitative analysis based on XRD data [14,36]. It is interesting to note that this deleterious effect was avoided using the transmission mode, as evidenced by the intensity ratio of the peaks d_{110} and d_{002} of the gibbsite (~50%). For comparison, the same samples were analyzed by reflection mode with manual sample holder assembling [14], resulting in a ratio d_{110}/d_{002} of only ~8%, a very low value considering the scale factor of this phase. At low angles, the noise is significant, although this mode of data collection allows a better resolution of possible peaks in this region of the diffractogram.

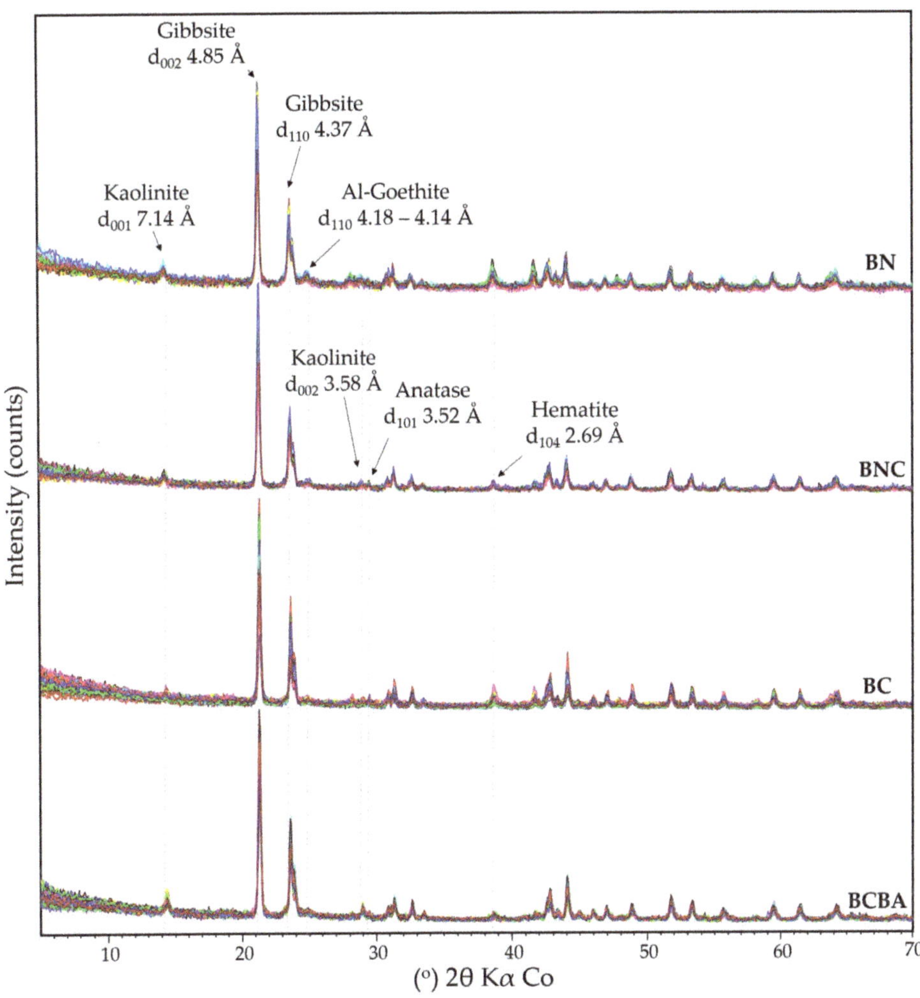

Figure 2. XRD patterns of the bauxite samples from BN, BNC, BC and BCBA lithologies.

3.2. Data Evaluation by Principal Component Analysis (PCA)

Figure 3 presents the score-plots for the first three principal components. As noted, the principal components PC-1, PC-2, and PC-3 explain, respectively, 35%, 22%, and 12% of the data variability (only 69% of the explained variance). Even considering 8 components, the explained variance remains lower than 75%. In contrast, Melo et al. [19] achieved 98% of the explained variance with only two components by using XRD reflection data. This shows that, although the preferred orientation effect was mitigated, the conditions of data collection by transmission mode used in this work resulted in a significant reduction in the intensities, which in turn, reduced the sensitivity of the statistical treatment in finding significant factors to reduce the data dimensionality.

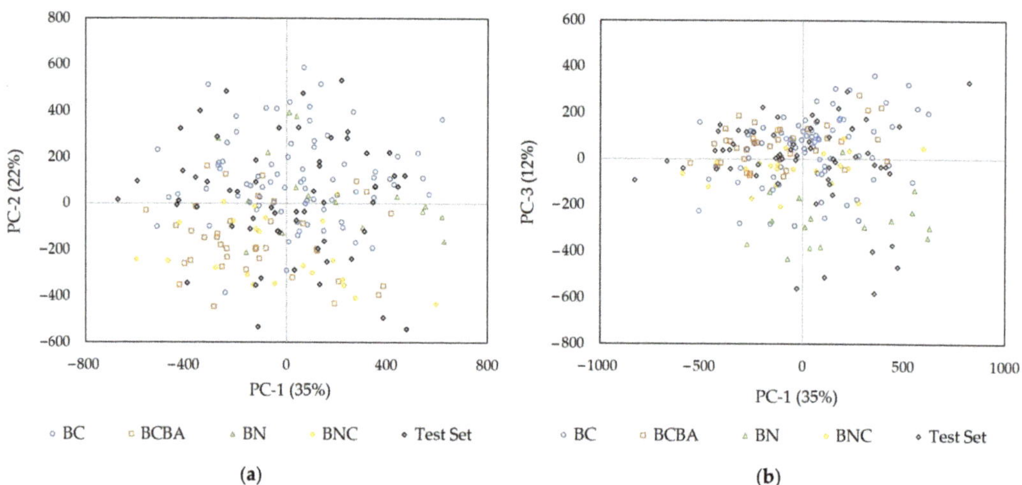

Figure 3. PCA score-plots: (a) PC-1 vs. PC-2; (b) PC-1 vs. PC-3. Explained variance of each component given in parentheses.

It is observed that no clustering is evidenced, even for those samples of the same lithology [8]. In this context, a K-means method was used to group the samples for further PLSR prediction. Figure 4 shows the PCA score-plot with the three clusters obtained (C1, C2, and C3). Although samples from the same lithology were classified into different clusters, C1 mostly contains BC; BNC and BCBA were mainly grouped in C1 and C2, while most BN samples were grouped in C3.

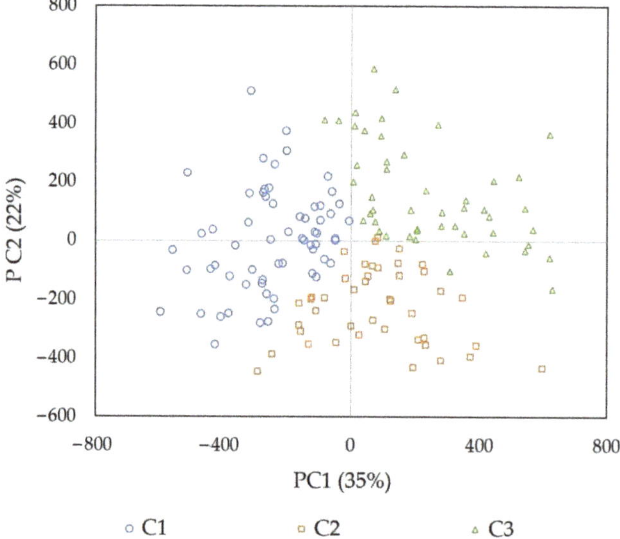

Figure 4. Clustering of samples by K-means represented in the PCA score-plot.

Figure 5 presents the linear-plot of the PCA loadings for PC-1, PC-2 (Figure 5a), and PC-6 (Figure 5b). It is observed that the reflections of the main mineral in the bauxite (gibbsite) have greater effect on PC-1 whilst PC-2 is more influenced by noise from low angles. The variability related to kaolinite were most extracted by PC-6.

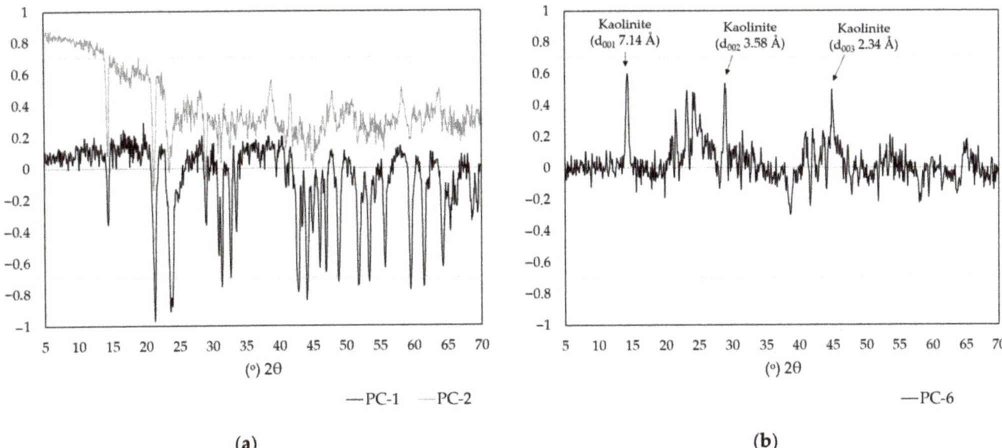

Figure 5. Linear-plot of PCA loadings: (**a**) PC-1 and PC-2; (**b**) PC-6. Dashed line represents 70% of the factors on the principal components.

3.3. Prediction of $AvAl_2O_3$ and $RxSiO_2$ by PLSR

Once the samples were classified into C1, C2, and C3 clusters, the respective sample calibration sets were used to build the PLSR models. Melo et al. [19] achieved an optimized condition of XRD data collection by increasing the step-size from 0.026° to 0.065° (2θ) and reducing the 2θ range up to 13–34°, such optimized condition for reflection mode allowed a less than 1 min XRD scan time. As observed in the PCA loadings-plot (Figure 5), using transmission data, the full pattern is relevant to extract the latent variables, therefore, in this study, the diffractograms were reduced only to the 13–65° (2θ) interval, just removing the background noise. This treatment resulted in a scan time of 1 min 15 s. Comparing to the traditional wet chemistry in which analyses can take 3–8 h, the use of PLSR on XRD data is much faster, being able to provide quick feedback to the process for decision-making.

After calibrating the models, defining the best pre-processing method for the dataset (mean-centered or standardized) and the number of factors to be included in the models through cross-validation, each sample in the test set was classified into one of the three clusters and then the parameters of bauxite quality control—$AvAl_2O_3$ and $RxSiO_2$ were predicted using the respective models. It can be observed in Figure 6 that there is a good fit of the predicted values for both parameters, mainly in those samples classified in C1.

Although the predicted mean values are close to the reference values, the models C2 and C3 showed a precision slightly lower than the acceptable limits for the quality control of bauxites [30]. The parameters that indicate the performance of the models are summarized in Tables 1 and 2 for $AvAl_2O_3$ and $RxSiO_2$, respectively.

A mean of residuals (mean of the difference between reference and predicted) close to zero denotes that the models present a good accuracy. The RMSEP denotes the precision of the model in the same unit as the predicted parameters (%$AvAl_2O_3$ and %$RxSiO_2$). Thus, a model with high precision presents lower RMSEP. In terms of bauxite quality control, a precision of <1.0–1.5% for $AvAl_2O_3$ and <0.5% for $RxSiO_2$ [14,25] is usually required. Feret [30] argues that these numbers are sometimes difficult to attain in the industrial practice using traditional wet chemistry methods.

The RPD indicates how well the model performs compared to using only the average of the original data [26]. Some authors argue that RPD < 1.0 denotes a very poor model, $1.0 \leq RPD < 1.4$ a poor model, $1.4 \leq RPD < 1.8$ a fair model, $1.8 \leq RPD < 2.0$ a good model and $RPD \geq 2.0$ an excellent model [26–28]. It can be observed that the models C1 and C2 for both $AvAl_2O_3$ and $RxSiO_2$ performed well with RPD ~2.0. It is interesting to note that, although the RMSEP of C3 model is high, it presented RPD = 2.9 which means an excellent model to predict $AvAl_2O_3$, denoting that the samples in this cluster have a wide range of $AvAl_2O_3$ content (min: 38.57%, max: 52.79%), and therefore, the model is sensitive to variations and capable of predicting this parameter satisfactorily.

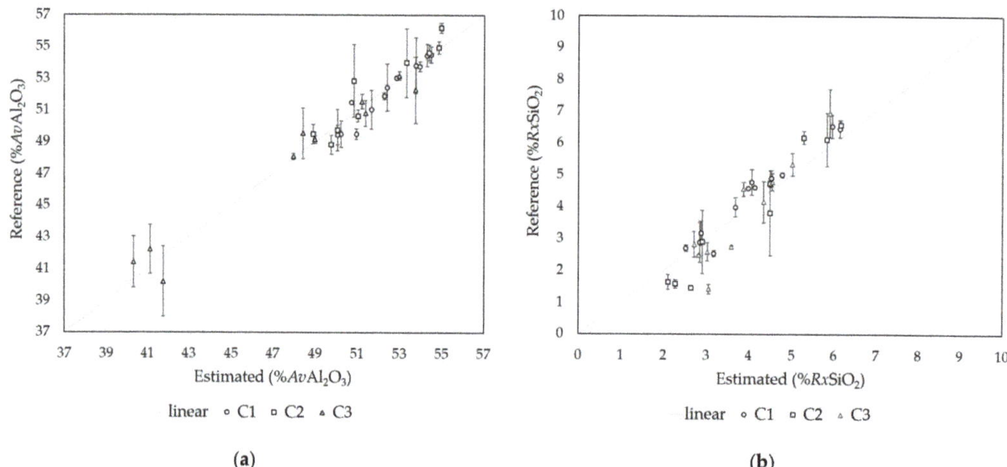

Figure 6. Comparison of reference and estimated values using PLSR models for each cluster: (**a**) $AvAl_2O_3$; (**b**) $RxSiO_2$. Bars indicating residual standard deviation (RSD).

Table 1. Results of the prediction of $AvAl_2O_3$ with the models for each cluster.

Models	Mean of Residuals	RMSEP (%)	RE (%)	RPD	Preprocessing Method (No. of Factors in the Model)
C1	0.24	0.85	1.63	1.98	Mean-centered (5)
C2	0.35	1.24	2.37	1.98	Mean-centered (4)
C3	−0.30	1.71	3.58	2.90	Mean-centered (5)

Table 2. Results of the prediction of $RxSiO_2$ with the models for each cluster.

Models	Mean of Residuals	RMSEP (%)	RE (%)	RPD	Preprocessing Method (No. of Factors in the Model)
C1	−0.28	0.49	10.72	2.38	Standardized (3)
C2	−0.14	0.78	17.90	2.04	Mean-centered (5)
C3	−0.11	0.78	19.16	1.36	Standardized (4)

It is interesting to note that the best model (C1) mostly contains samples of the main ore lithology (BC), denoting that the method is suitable for quality control. In contrast, the worst model (C3) is mainly related to the samples of the marginal ore (BN, generally considered as gangue). This bauxite lithology presents the highest kaolinite content and the lowest gibbsite content in the Miltonia 3 bauxite profile. The low degree of crystallinity of the kaolinite in this lithology affects the XRD profile [13,14]. Melo et al. [19] also observed a lower precision for this material that could be related to the preferred orientation. As this effect was eliminated using transmission mode, the low precision of the C3 model may be related to the wide range of $AvAl_2O_3$ and $RxSiO_2$ content in the sample group or another unknown effect. It also may represent a limitation of this method (or an optimization point), that is, it is not suitable for geological survey applications; where it may be present, samples that strongly deviate from the ore in terms of mineralogical composition and phases' content.

Feret et al. [37] state that methods of phase quantification based on regression may successfully be used in bauxite exploration, however, they are deposit-specific. Similarly, the method presented in this work (and also in Melo et al. [19]) represents a case study with lateritic gibbsitic bauxite from the Miltonia plateau. Although König and Norberg [18] reported satisfactory prediction using a generic PLSR model for bauxites from different locations, the results suggest that changes in the ore's mineralogy, related only to the concentration and crystallinity of the phases, may impact prediction. Nevertheless, it is believed that the methodology can be easily adapted to bauxites of different origins, in particular, the Amazon lateritic bauxites (Paragominas plateau Miltonia 5, Juruti, Trombetas and Rondon do Pará) with similar mineralogy and containing only gibbsite as an aluminum-bearing mineral [6,12,16,38].

As depicted in Figure 7, the coefficients and factors of each PLSR model can be plotted in relation to the °2θ position, allowing to interpret the obtained model in terms of the XRD pattern of the clustered bauxite samples. It is observed that the largest coefficients of the $AvAl_2O_3$ models (C1, C2, and C3) are negatively correlated with kaolinite basal reflections and positively correlated with gibbsite (d_{110}), whereas basal gibbsite reflection (d_{002}) has greater weights on the factor loadings. The asymmetry shape of this reflection, however, was revealed as at least two generations of gibbsites. Similar results were observed by Melo et al. [19], and according to König et al. [10] and Negrão et al. [16], it could be associated with aluminum-rich horizons, where along the bauxite profile, well-formed coarse gibbsite crystals fills microvoids, and a new generation of fine, poor-crystalline gibbsite is dispersed in the matrix. Interestingly, Al-goethite (d_{110}) has a high impact on model coefficients (highlighted area in Figure 7a,c). The respective broad peak area in C1 and C2 coefficients denote the presence of several %Al-substitution in the goethite structure. It is also noted that, in the C3 model, a wide area from 15–20° 2θ presented high coefficients. This area, however, has no XRD reflection associated, and therefore, may be related to the amorphous in these samples. It was not possible to quantify the amorphous, but it is assumed that this is more evident in the overlying lithologies, in particular, BNC and BN due to different laterization cycles and the greater presence of neoformed minerals [8,16]. This assumption is in agreement with the results since BN samples were mainly grouped in C3, and is probably related to the relatively low prediction of this model.

On the models for $RxSiO_2$, higher coefficients related to kaolinite and Al-goethite were also observed, however, positively. For models with standardized datasets (C1 and C3), the basal reflections of kaolinite have greater impact on loadings, while in the C2 model (mean-centered), the d_{002} peak of gibbsite had greater weight. A double peak referring to the d001 of kaolinite was observed for the coefficients of models C2 and C3, which must also be associated with kaolinite generations with different degrees of crystallinity. Melo et al. [8] demonstrated that this difference actually occurs, so that kaolinites from the overlying lithologies are less ordered and, consequently, more reactive to the Bayer process. It is interesting to note that, although this significantly impacts processing costs, this information is not known by the industry in the context where all the quality control of bauxite is by traditional wet methods. The variability of kaolinite crystallinity in these two clusters may be associated with the reduced accuracy of the respective models.

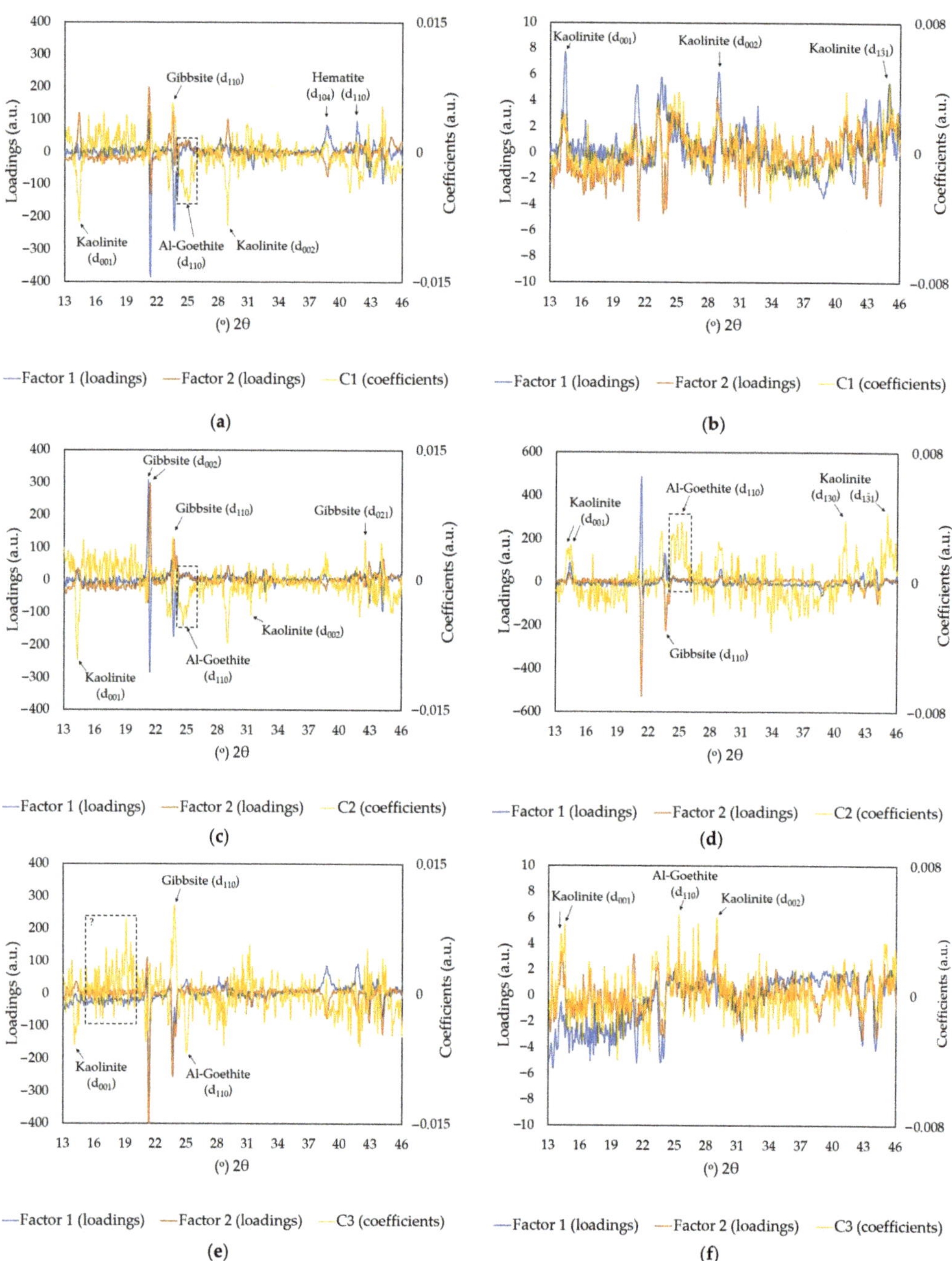

Figure 7. Linear representation of the coefficients and main factors of the models: (**a**) AvAl$_2$O$_3$ (C1); (**b**) RxSiO$_2$ (C1); (**c**) AvAl$_2$O$_3$ (C2); (**d**) RxSiO$_2$ (C2); (**e**) AvAl$_2$O$_3$ (C3), and (**f**) RxSiO$_2$ (C3).

4. Conclusions

In this work, the use of a method for quality control of bauxites based on statistical tools on XRD transmission data was evaluated. The method comprises classifying bauxite samples by PCA and K-means according to their latent mineralogical characteristics, building PLSR models for each sample group, and using these models to predict the $AvAl_2O_3$ and $RxSiO_2$ parameters in new samples.

The samples were classified into three clusters (C1, C2, and C3) and the respective models were evaluated in relation to the wet chemistry reference values. The C1 model presented satisfactory accuracy and precision for both parameters. The RMSEP of 0.85% ($AvAl_2O_3$) and 0.49% ($RxSiO_2$) attain the required limits (1.0–1.5% and 0.5%, respectively). The C2 and C3 models, related to marginal ore lithologies presented satisfactory accuracy but low precision.

These results also indicate that, although the preferred orientation was eliminated using the XRD transmission data collection, there was no incremental improvement compared with the PLSR models obtained with reflection data [14].

These results cleared showed that the methodology can be applied for quality control in the beneficiation plant, but not suitable for geological survey applications. It is worth mentioning that this method presents several advantages over traditional wet chemistry, mainly due to its speed (less than 5 min to run XRD analysis and obtain the prediction), ease of being completely automated, and no dangerous chemical reagents are required.

Author Contributions: Conceptualization, C.C.A.M.; methodology, C.C.A.M. and R.S.A.; formal analysis, C.C.A.M.; resources, data curation, S.P.A.P.; writing—original draft preparation, C.C.A.M. writing—review and editing, S.P.A.P. and R.S.A.; supervision, project administration, and funding acquisition, S.P.A.P. All authors have read and agreed to the published version of the manuscript.

Funding: This research was funded by CNPq (Conselho Nacional de Desenvolvimento Científico e Tecnológico), grant number 402.427/2016-5 (Edital Universal, for the following project: "Desenvolvimento e Otimização de Métodos Prospectivos e de Controle de Beneficiamento de Bauxitas Gibbsíticas").

Data Availability Statement: The data presented in this study is contained within the present article.

Acknowledgments: The authors thank the Brazilian agencies: CAPES (Coordenação de Aperfeiçoamento de Pessoal de Nível Superior) for a PhD scholarship to the first author and CNPq (Conselho Nacional de Desenvolvimento Científico e Tecnológico). We also acknowledge the Mineração Paragominas SA Company (Norsk Hydro) for the support.

Conflicts of Interest: The authors declare no conflict of interest.

References

1. US Geological Survey. *Mineral Commodity Summaries 2021*; Government Printing Office: Washinton, DC, USA, 2021; p. 200.
2. Paz, S.; Angélica, R.; Kahn, H. Optimization of the reactive silica quantification method applied to Paragominas-type gibbsitic bauxites. *Int. J. Miner. Process.* **2017**, *162*, 48–57. [CrossRef]
3. Authier-Martin, M.; Forte, G.; Ostap, S.; See, J. The mineralogy of bauxite for producing smelter-grade alumina. *JOM* **2001**, *53*, 36–40. [CrossRef]
4. Ostap, S. Control of silica in the Bayer process used for alumina production. *Can. Metall. Q.* **1986**, *25*, 101–106. [CrossRef]
5. Smith, P. The processing of high silica bauxites—review of existing and potential processes. *Hydrometallurgy* **2009**, *98*, 162–176. [CrossRef]
6. Angélica, R.S.; Kahn, H.; Paz, S. A proposal for bauxite quality control using the combined Rietveld-Le Bail-Internal Standard PXRD method–Part 2: Application to a gibbsitic bauxite from the Paragominas region, northern Brazil. *Miner. Eng.* **2018**, *122*, 148–155. [CrossRef]
7. Bredell, J.H. Calculation of available alumina in bauxite during reconnaissance exploration. *Econ. Geol.* **1983**, *78*, 319–325. [CrossRef]
8. Melo, C.C.A.; Oliveira, K.S.; Angélica, R.S.; Paz, S. Análise de agrupamentos associada a difratometria de raios-X: Uma classificação mineralógica prática de bauxitas e seus produtos de digestão Bayer. *Holos* **2017**, *6*, 32. [CrossRef]
9. O'Connor, D.J. *Alumina Extraction from Nom Bauxitic Materials*; Aluminium-Verlag: Sydney, Australia, 1988; p. 370.
10. König, U.; Angélica, R.S.; Norberg, N.; Gobbo, L. Rapid X-ray diffraction (XRD) for grade control of bauxites. *ICSOBA Proc.* **2012**, *19*, 11.
11. Paz, S.; Kahn, H.; Angélica, R.S. A proposal for bauxite quality control using the combined Rietveld–Le Bail–Internal Standard PXRD Method–Part 1: Hkl model developed for kaolinite. *Miner. Eng.* **2018**, *118*, 52–61. [CrossRef]

12. Neumann, R.; Avelar, A.N.; Da Costa, G.M. Refinement of the isomorphic substitutions in goethite and hematite by the Rietveld method, and relevance to bauxite characterisation and processing. *Miner. Eng.* **2014**, *55*, 80–86. [CrossRef]

13. Feret, F.R. Selected applications of Rietveld-XRD analysis for raw materials of the aluminum industry. *Powder Diffr.* **2013**, *28*, 112–123. [CrossRef]

14. Aylmore, M.G.; Walker, G.S. The quantification of lateritic bauxite minerals using X-ray powder diffraction by the Rietveld method. *Powder Diffr.* **1998**, *13*, 136–143. [CrossRef]

15. Nong, L.; Yang, X.; Zeng, L.; Liu, J. Qualitative and quantitative phase analyses of Pingguo bauxite mineral using X-ray powder diffraction and the Rietveld method. *Powder Diffr.* **2007**, *22*, 300–302. [CrossRef]

16. Negrao, L.; Da Costa, M.L.; Pöllmann, H.; Horn, A. An application of the Rietveld refinement method to the mineralogy of a bauxite-bearing regolith in the Lower Amazon. *Miner. Mag.* **2018**, *82*, 413–431. [CrossRef]

17. Feret, F.; See, J. Analysis of bauxite by X-ray diffraction using synchrotron radiation. In Proceedings of the ICSOBA Conference, Zhengzhou, China, 25–28 November 2010.

18. König, U.; Norberg, N. New Tools for Process Control in Aluminium Industries-PLSR on XRD Raw Data. In AQW Proceedings 23 March 2019. Available online: http://www.aqw.com.au/papers/item/new-tools-for-process-control-in-aluminium-industries-plsr-on-xrd-raw-data.html (accessed on 5 May 2021).

19. Melo, C.C.A.; Angélica, R.S.; Paz, S.P.A. A proposal for rapid grade control of gibbsitic bauxites using multivariate statistics on XRD data. *Miner. Eng.* **2020**, *157*, 106539. [CrossRef]

20. Melo, C.C.A.; Paz, S.P.A. Estimation of AvAl$_2$O$_3$ and RxSiO$_2$ by Partial Least Square Regression (PLSR) on XRD data: A case study using low grade bauxites. In Proceedings of the 36th International ICSOBA Conference, Belem, Brazil, 29 October–1 November 2018.

21. Carneiro, C.D.C.; Yanez, D.N.D.V.S.; Ulsen, C.; Fraser, S.J.; Antoniassi, J.L.; Paz, S.; Angelica, R.S.; Kahn, H. Imputation of reactive silica and available alumina in bauxites by self-organizing maps. In Proceedings of the 2017 12th International Workshop on Self-Organizing Maps and Learning Vector Quantization, Clustering and Data Visualization (WSOM), Nancy, France, 28–30 June 2017.

22. Wold, S.; Sjöström, M.; Eriksson, L. PLS-regression: A basic tool of chemometrics. *Chemometr. Intell. Lab. Syst.* **2001**, *58*, 109–130. [CrossRef]

23. Bro, R.; Smilde, A.K. Principal component analysis. *Anal. Methods* **2014**, *6*, 2812–2831. [CrossRef]

24. Cordella, C.B.Y. PCA: The basic building block of chemometrics. *Anal. Chem.* **2012**, *7*. [CrossRef]

25. Rossel, R.V.; Walvoort, D.; McBratney, A.; Janik, L.; Skjemstad, J. Visible, near infrared, mid infrared or combined diffuse reflectance spectroscopy for simultaneous assessment of various soil properties. *Geoderma* **2006**, *131*, 59–75. [CrossRef]

26. Rossel, R.V.; McGlynn, R.; McBratney, A. Determining the composition of mineral-organic mixes using UV–vis–NIR diffuse reflectance spectroscopy. *Geoderma* **2006**, *137*, 70–82. [CrossRef]

27. Olatunde, K.A. Determination of petroleum hydrocarbon contamination in soil using VNIRDRS and PLSR modeling. *Heliyon* **2021**, *7*, e067942. [CrossRef] [PubMed]

28. dos Santos, F.R.; de Oliveira, J.F.; Bona, E.; dos Santos, J.V.F.; Barboza, G.M.; Melquiades, F. EDXRF spectral data combined with PLSR to determine some soil fertility indicators. *Microchem. J.* **2020**, *152*, 104275. [CrossRef]

29. König, U.; Degen, T.; Norberg, N. PLSR as new XRD method for downstream processing of ores–case study: Fe^{2+} determina-tion in iron ore sinter. *Powder Diffr.* **2014**, *29*, 78–83. [CrossRef]

30. Feret, F.R. Inaccuracies in Estimation of Bauxite Extractable and Mineralogical Constituents. In Proceedings of the 36th International ICSOBA Conference, Belem, Brazil, 29 October–1 November 2018.

31. Picanço, F.E.; Pinheiro, D.C.; Calado, W.M.; Almeida, K.R.; Quadros, C.R.R.; Rabelo, S.A. Amostragem em diferentes malhas de sondagem para minério de bauxita—Mina de Miltonia 03—Paragominas-PA. In Proceedings of the Anais do 13° Simpósio de Geologia da Amazônia, Belem, Brazil, 22–26 September 2013.

32. Silva, H.M.; Picanço, E.; Maurity, C.; Morais, W.; Santos, H.C.; Guimarães, O. Geology, mining operation and scheduling of the Paragominas bauxite mine. In Proceedings of the 8th International Alumina Quality Workshop, 7–12 September 2008; The Workshop: Darwin, NT, Australia, 2008; pp. 11–16.

33. Geladi, P.; Kowalski, B.R. Partial least-squares regression: A tutorial. *Anal. Chim. Acta* **1986**, *185*, 1–17. [CrossRef]

34. Paz, S.P.A.; Torres, P.W.T.S.; Angélica, R.S.; Kahn, H. Synthesis, Rietveld refinement and DSC analysis of Al-goethites to support mineralogical quantification of gibbsitic bauxites. *J. Therm. Anal. Calorim.* **2017**, *128*, 841–854. [CrossRef]

35. Moore, D.M.; Reynolds, R.C. *X-ray Diffraction and the Identification and Analysis of Clay Minerals*, 2nd ed.; Oxford University Press: Oxford, UK, 1997.

36. Zhou, X.; Liu, D.; Bu, H.; Deng, L.; Liu, H.; Yuan, P.; Du, P.; Song, H. XRD-based quantitative analysis of clay minerals using reference intensity ratios, mineral intensity factors, Rietveld, and full pattern summation methods: A critical review. *Solid Earth Sci.* **2018**, *3*, 16–29. [CrossRef]

37. Feret, F.; Authier-Martin, M.; Sajó, I. Quantitative phase analysis of Bidi-Koum bauxites (Guinea). *Clay Clay Miner.* **1997**, *45*, 418–427. [CrossRef]

38. Negrão, L.; Pöllmann, H.; Alves, T.C. Mineralogical Appraisal of Bauxite Overburdens from Brazil. *Minerals* **2021**, *11*, 677. [CrossRef]

Article

Heavy Mineral Sands Mining and Downstream Processing: Value of Mineralogical Monitoring Using XRD

Uwe König [1,*] and Sabine M. C. Verryn [2]

1   Malvern Panalytical B.V., Lelyweg 1, 7602 EA Almelo, The Netherlands
2   XRD Analytical and Consulting, 75 Kafue Street, Pretoria 0081, South Africa; sabine.verryn@xrd.co.za
*   Correspondence: uwe.koenig@malvernpanalytical.com

Abstract: Heavy mineral sands are the source of various commodities such as white titanium dioxide pigment and titanium metal. The three case studies in this paper show the value of X-ray diffraction (XRD) and statistical methods such as data clustering for process optimization and quality control during heavy mineral processing. The potential of XRD as an automatable, reliable tool, useful in the characterization of heavy mineral concentrates, product streams and titania slag is demonstrated. The recent development of ultra-high-speed X-ray detectors and automated quantification allows for 'on the fly' quantitative X-ray diffraction analysis and truly interactive process control, especially in the sector of heavy mineral concentration and processing. Apart from the information about the composition of a raw ore, heavy mineral concentrate and the various product streams or titania slag, this paper provides useful information by the quantitative determination of the crystalline phases and the amorphous content. The analysis of the phases can help to optimize the concentration of ores and reduction of ilmenite concentrate. Traditionally, quality control of heavy mineral concentrates and titania slag relies mainly on elemental, chemical, gravimetrical, and magnetic analysis. Since the efficiency of concentration of minerals in the different product streams and reduction depends on the content of the different minerals, and for the latter on the titanium and iron phases such as ilmenite $FeTiO_3$, rutile TiO_2, anatase TiO_2, or the various titanium oxides with different oxidation stages, fast and direct analysis of the phases is required.

Keywords: heavy minerals; ilmenite; titania slag; XRD; cluster analysis; rietveld; Magneli phases

Citation: König, U.; Verryn, S.M.C. Heavy Mineral Sands Mining and Downstream Processing: Value of Mineralogical Monitoring Using XRD. *Minerals* **2021**, *11*, 1253. https://doi.org/10.3390/min11111253

Academic Editor: Mark I. Pownceby

Received: 4 October 2021
Accepted: 4 November 2021
Published: 11 November 2021

1. Introduction

Mineral sands, also commonly known as heavy minerals because of their relatively high specific density compared to sand (more than 4 kg/m^3), are an important source of various titanium raw materials and zircon. These metals and their oxides have become extremely important for various sectors of industry.

Heavy minerals naturally occur in relatively small concentrations, so that specialized preparation systems equipped with spiral separators, electrostatic separators and magnetic separators are required for concentration and separation.

Mineral beneficiation relies on quantitative knowledge of phases present in the process streams. This information is necessary to optimize each step from the flowsheet in Figure 1 to process and upgrade the mined heavy mineral sand. Often X-ray fluorescence (XRF) or wet chemistry are used to obtain elemental information which assists the metallurgists to estimate the mineral processing performance. When more accurate mineral phase information is needed, X-ray diffraction (XRD) is the most useful technique as a fast and accurate alternative to time consuming scanning electron microscopy. Due to the diverse applications and the improvements in speed, accuracy, and flexibility of the analysis, during the last two decades, XRD has become a standard tool in industries, such as cement or aluminium production [1–3].

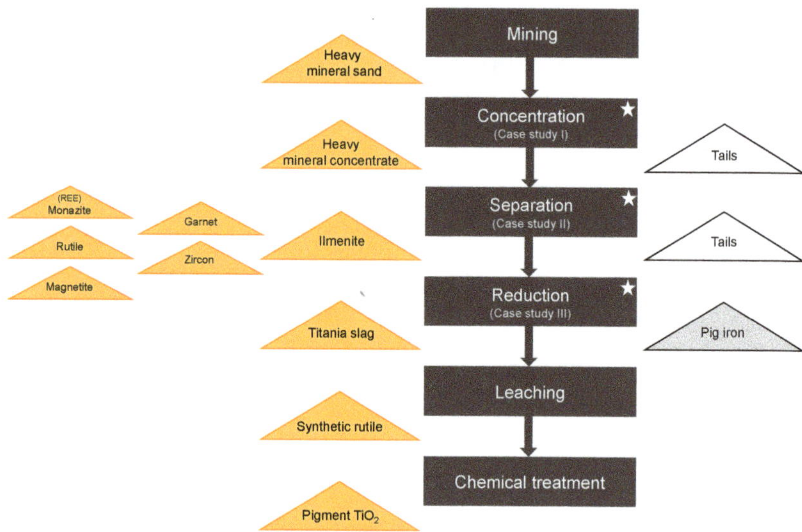

Figure 1. Simplified flowsheet for the process of heavy mineral separation and TiO$_2$ production. Left triangles = intermediate and final products, middle = process flow, right triangles = by-products or waste. Stars indicate the case studies investigated in this paper.

Figure 1 simplifies a typical process flow for the extraction of TiO$_2$ from heavy mineral sands, as well as further products of interest. Separation of the valuable heavy minerals from the primary ore is carried out in two stages: (a) wet concentration, utilizing sizing and gravity differentiation between heavy minerals, clay and quartz, and (b) dry separation, exploiting the magnetic and electrostatic properties of the minerals of interest. Case studies I and II deal with the fast analysis of heavy mineral concentrates improving the efficiency of the separation processes. Case study III demonstrates the use of XRD for checking the performance of the ilmenite smelting (reduction) process.

Ilmenite as well as rutile are the principal feedstocks for titanium dioxide production. Titanium dioxide is used as pigment for the manufacture of paints, coatings and plastics and also in other applications such as pharmaceutical, paper, cosmetics (e.g., sunscreen), toothpaste, inks and fibres. Titanium dioxide can be used in specialist applications including welding rods and the production of titanium metal for industrial and aerospace applications.

Zircon is used in the manufacture of ceramic products including tiles, sanitaryware and tableware, and as an opacifier in surface glazes and pigments. Zircon is also the main component in the production of zirconium chemicals used in antiperspirants, paper coatings, paint driers and catalysts. Zircon can also be used as foundry sand, cathode ray tube television glass and in refractories.

Most garnet is mined for industrial uses such waterjet cutting purposes, abrasive blasting media, water filtration granules and abrasive powders.

The objective of the three case studies in this paper is to demonstrate that X-ray diffraction (XRD) can be used as fast and reliable process control tool for heavy mineral sand concentration and beneficiation.

2. Materials and Methods

The three case studies form part of the typical process flow for the extraction of TiO$_2$ from heavy mineral sands. Case study I describes the monitoring of the concentration of heavy mineral sands using statistical cluster analysis. Case study II demonstrates the mineralogical quantification of the different material streams during separation of the various heavy minerals. The use of XRD for process control during ilmenite smelting is discussed in case study III.

2.1. Samples and Sample Preparation

For case studies I and III, samples were prepared as pressed pellets using automatic sample preparation equipment for minimized preferred orientation and to guarantee a constant sample preparation quality. All powder samples were milled for 30 s and pressed for 30 s at 10 tons into steel ring sample holders. Constant sample preparation and the highest sample throughput could be realized using fully automated preparation and analysis setups.

For case study II samples were prepared using a backloading preparation technique. These samples were milled for 10 min in a Cr-steel milling vessel as it was found to be the optimum time for these Heavy Mineral Concentrate (HMC) samples.

2.2. X-ray Diffraction (XRD)

X-ray diffraction (XRD) is a versatile, nondestructive analytical method for identification and quantitative determination of crystalline phases present in powdered and bulk samples. Establishing which phases are present in a sample is usually the first step of a whole series of analyses and forms the basis of investigations on how much of each phase is present (quantitative phase analysis).

All crystalline materials have their own unique, characteristic X-ray fingerprint based on their crystal structure. When diffraction data for a particular sample are compared against a database of known materials, the crystalline phases within the sample can be identified. For this study the Crystallography Open Database (COD) was used [4]. Data evaluation was performed with the software package HighScore Plus version 4.9 [5]. Diffractograms were measured with a Malvern Panalytical Aeris Minerals diffractometer with a PIXcel detector, fixed slits and Fe-filtered Co-Kα radiation. The instrument was optimized for the rough needs of mineral industry environments (remote control possible), featuring measurement times of 5–10 min per scan. The samples were measured under room temperature within a range of 10 to 80°2θ and a step size of 0.02°2θ. Compared with traditional XRD approaches, the capabilities of high-speed XRD detectors allow measurements within minutes instead of hours.

2.3. Rietveld Quantification

The mineral quantification of all samples was determined using the Rietveld method [6–8]. Modern XRD quantification analysis techniques such as Rietveld analysis are attractive alternatives to classical peak intensity or area-based methods since they do not require any standards or monitors. The Rietveld method offers impressive accuracy and speed of analysis. The knowledge of the exact crystal structure of all minerals present in the samples is mandatory for the Rietveld refinements. An example of the accuracy of the results is given in Section 3.2, using a comparison of TiO_2 content recalculated from Rietveld refinement results with TiO_2 content determined by X-Ray Fluorescence Spectroscopy (XRF). A quality indicator of a Rietveld refinement is the weighted profile R-value (R_{wp}), which is minimized. The weighting is such that higher intensity data is more important than lower intensity data. Therefore, fitting the peaks is more important than fitting the background. In general, this value should be less than 10, [8]. Amounts below 0.5 w%, may, however, be unreliable. The advantage of using Rietveld analysis over automated quantitative scanning electron microscope analysis for the analysis of HMC samples, such as speed of analysis, sample size, and cost effectiveness, is demonstrated in [2].

2.4. Cluster Analysis

To handle large amounts of data achieved by rapid data collection using a linear detector, "cluster analysis" is a useful tool to group different XRD measurements into similar clusters, [9,10]. The method can be used to sort different ore grades with different mineralogical composition and thus varying process behavior or to detect instabilities during processing.

Cluster analysis greatly simplifies the analysis of large amounts of data. It automatically sorts all (closely related) scans of an experiment into separate groups and marks the most representative scan of each group as well as the most outlying scans within each group. Cluster analysis is basically a three-step process, but it contains optional visualization and verifications steps as well:

1. Comparison of all scans in a document with each other. The result is a correlation matrix representing the distances (or dissimilarity) of all data points of any given pair of scans.

2. Agglomerative hierarchical cluster analysis puts the scans in different classes defined by their similarity. The output of this step is displayed as a dendrogram, where each scan starts at the left side as an individual cluster. The clusters amalgamate in a stepwise fashion until they are all united in one single group.

3. The best possible grouping (=number of separate clusters) is estimated by the KGS test, named after Kelley, Gardner, Sutcliffe [9], or by the largest relative step on the dissimilarity scale. Additionally, the most representative scan and the two most outlying scans within each cluster are determined and marked.

4. As well as hierarchical clustering, independent tools such as Principal Components Analysis (PCA) can be used to define clusters. The PCA method finds systematic large variances in the set of observations, i.e., the so-called principal components or "eigenvalues". It uses the correlation matrix as input and displays the three most important principal components in a pseudo-3D plot. They explain the major part of the total overall variance in the correlation matrix.

3. Results

3.1. Case Study I: Cluster Analysis of Heavy Minerals Concentrates

For this case study, cluster analysis was used to classify heavy mineral concentrates from different spiral stages of the concentration process at a wet concentration plant. Gravity separation with spirals was performed after the ore had passed scrubbing and screening. The wet concentration process consists of several stages: rougher spiral, cleaner spiral, scavenger spirals, and re-cleaner spirals, as shown in Figure 2. Intermediate heavy minerals concentrate from each spiral stage passes to the next stage to be further concentrated.

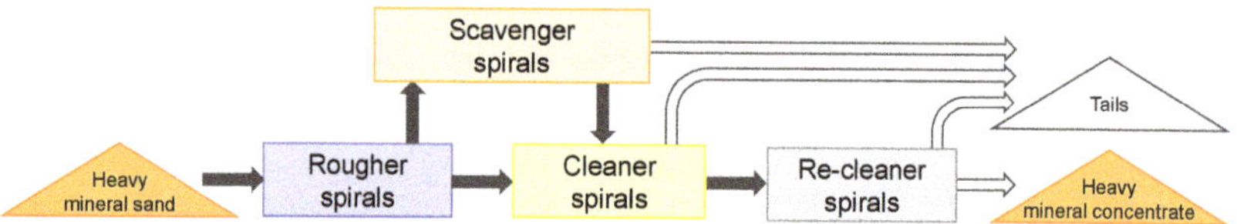

Figure 2. Simplified flow sheet of spiral separation of heavy minerals.

The XRD pattern of 46 samples from the four different spiral stages were collected. Without further processing, a cluster analysis was performed on the measurement. The principal component analysis (PCA) score plot in Figure 3 shows four groups of samples.

The four clusters represent the four different spirals. The spread of the data points indicates the diversity in the mineralogical phase content in the clusters. Samples from the rougher spirals have a more diverse mineralogical composition compared to the other three clusters. Most similar are the samples coming from the re-cleaner spirals.

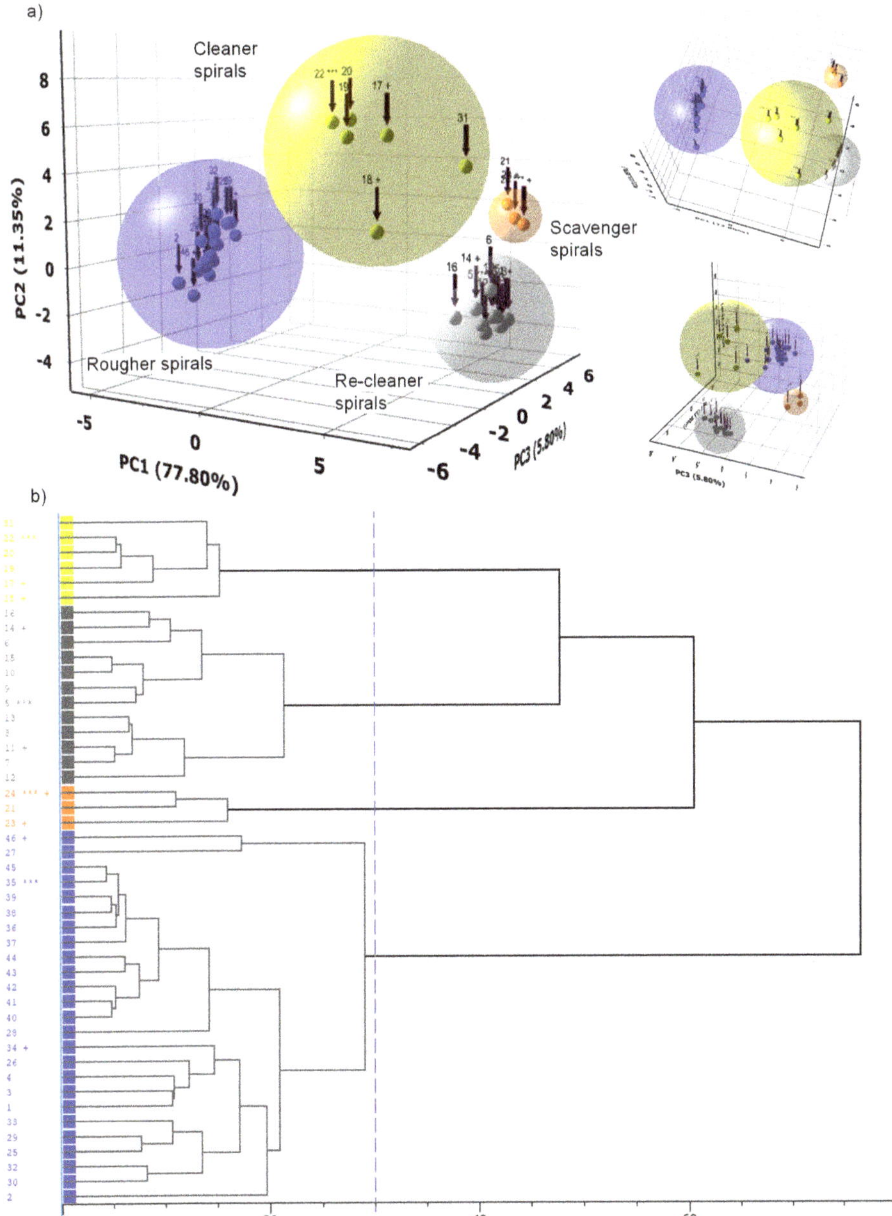

Figure 3. (**a**) 3D PCA score plot for the cluster analysis of 46 samples from a wet concentration plant, illustrated in three views. The three axes represent the three most important principal components explaining a major part of the total overall variance in the correlation matrix. The covered total variance is 94.95%. (**b**) Corresponding dendrogram with agglomerative hierarchical cluster analysis showing the scans in different classes defined by their similarity and vertical cut-off lines. Each dataset is connected by a tie bar to another dataset or to a cluster. The horizontal length of the tie bars is proportional to the dissimilarity (*x*-axis). The color of the spheres indicates the cluster they belong to. The *** after a scan name indicates the most representative scan of a cluster. The + after a scan name indicates the two most different (outlying) scans within one cluster. Each dot represents one measurement. All measurements are compared with each other to find similarities.

Generally, a larger spread within the cluster illustrates either varying compositions of heavy mineral concentrates or unstable separation conditions within the spirals. Outliers, which do not belong to any cluster, usually signal problems in the processing step. As a result, cluster analysis can be used as a fast tool to immediately control and adjust the efficiency and accuracy of the spiral stages. This enables fast responses to changes in raw materials or process conditions and with that, a greater focus on profitability. Cluster analysis not only allows producers to standardize and monitor concentration processes objectively, but it enables them to compare spiral performance across different processing plants quickly and easily.

Cluster analysis can be performed before further investigations on measurements, such as phase identification and quantification. The most representative scan and the scans that differ most within a cluster can be used as starting points for more detailed investigations.

3.2. Case Study II: Quantification of Heavy Minerals Concentrates

Samples for this case study originated from a heavy mineral sand mine in South Africa. The Heavy Minerals Concentrate (HMC) was split into three commercial products through various processes in the mineral processing plant. There were three concentrate product streams: ilmenite concentrate, high grade zircon/rutile and high-grade garnet concentrate (almandine).

This study focuses on these three streams as they are the main commodities for the mine. Minor additional products such as magnetite concentrates are also marketed. REE minerals only occur as traces and are currently not commercialized. Heavy mineral sands are mined in a free dig operation using conventional trucking and excavation using mobile excavators, front-end loaders and trucks. The ore is processed via a Primary Bulk Concentrator (PBC), where the minerals are separated by a chemical-free gravity process. Subsequently garnet is separated at a garnet stripping plant and the nonmagnetic material is separated from the heavy mineral concentrate as shown in Figure 4.

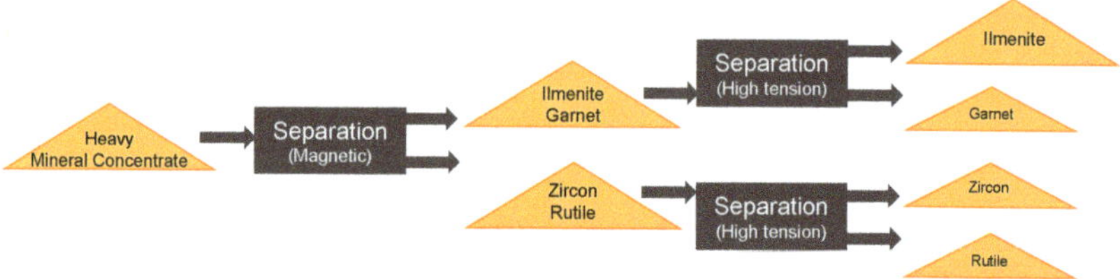

Figure 4. Simplified flow sheet of heavy mineral separation.

The HMC quantified in Figure 5 is the product of density separation of beach sand composites. The typical amount of total heavy minerals ranges between 10% and 45%.

After separation each of the product streams must fulfill predefined specifications not only based on chemical composition but also mineral quantities. Frequent monitoring of the mineral composition allows for estimation of stockpile compositions and optimal blending of materials to achieve the quality targets before shipment. XRD together with Rietveld quantification is a fast and easy method that can be completely automated. Figures 5–8 show examples of Rietveld quantifications of a typical heavy mineral concentrate (HMC), ilmenite, garnet and zircon concentrates. Each concentrate represents a weekly composite sample of 21 subsamples.

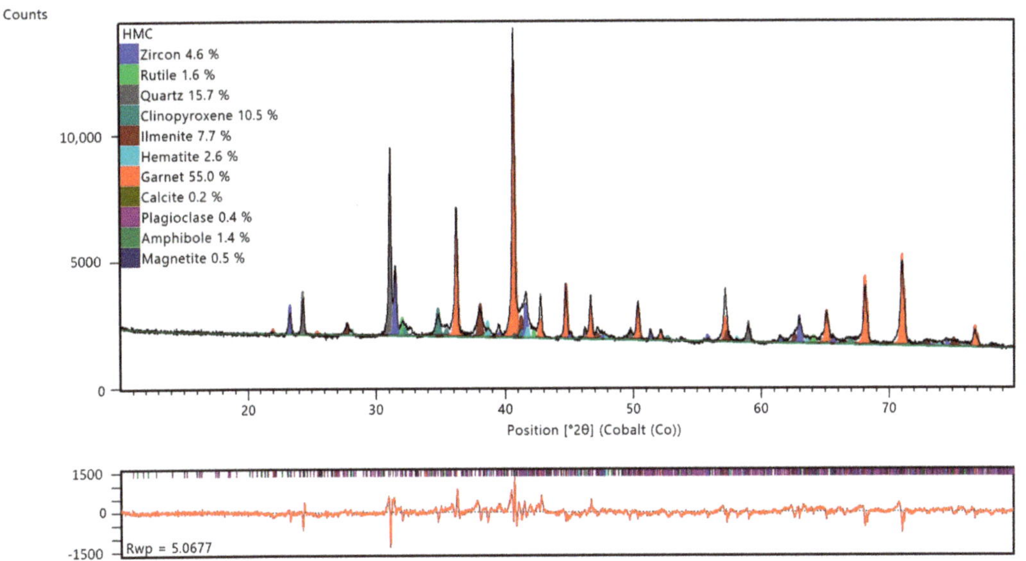

Figure 5. Typical Heavy Mineral Concentrate (HMC). Rietveld quantification (**top**) including the difference plot between measured and calculated profiles, as well as the R_{wp} (**bottom**).

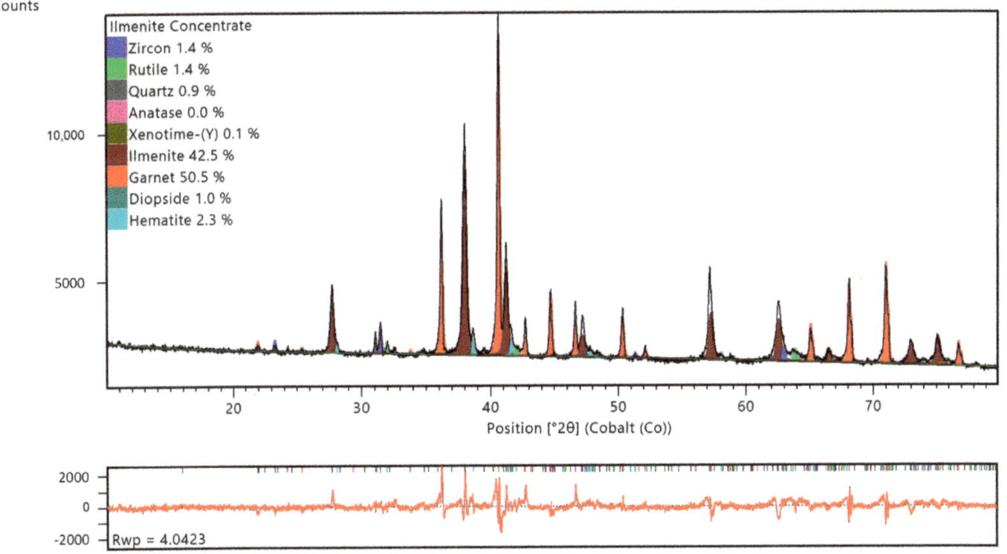

Figure 6. Typical ilmenite concentrate. Rietveld quantification (**top**) including difference plot between measured and calculated profiles, as well as the R_{wp} (**bottom**).

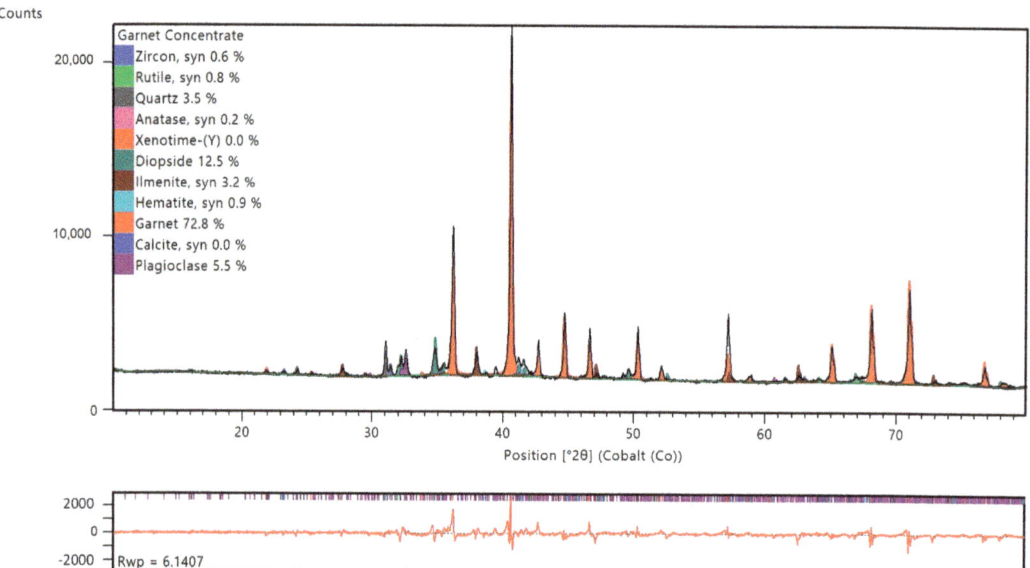

Figure 7. Typical garnet concentrate. Rietveld quantification (**top**) including difference plot between measured and calculated profiles as well as the R_{wp} (**bottom**).

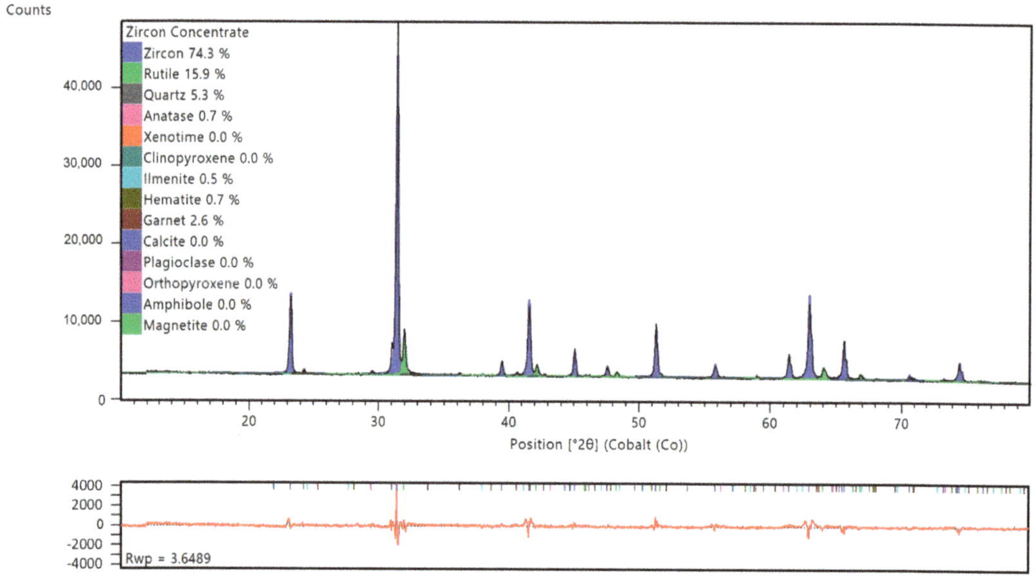

Figure 8. Typical zircon concentrate. Rietveld quantification (**top**) including difference plot between measured and calculated profiles as well as the R_{wp} (**bottom**).

Figure 9 shows the quantitative results of the heavy minerals content in the ilmenite product over several days. In this example of samples mainly containing ilmenite and garnet, the comparison of weight % TiO_2 between XRF analysis and TiO_2 content recalculated from Rietveld refinement results is a suitable indicator of the accuracy and relevance of using XRD for quantification of mineral content.

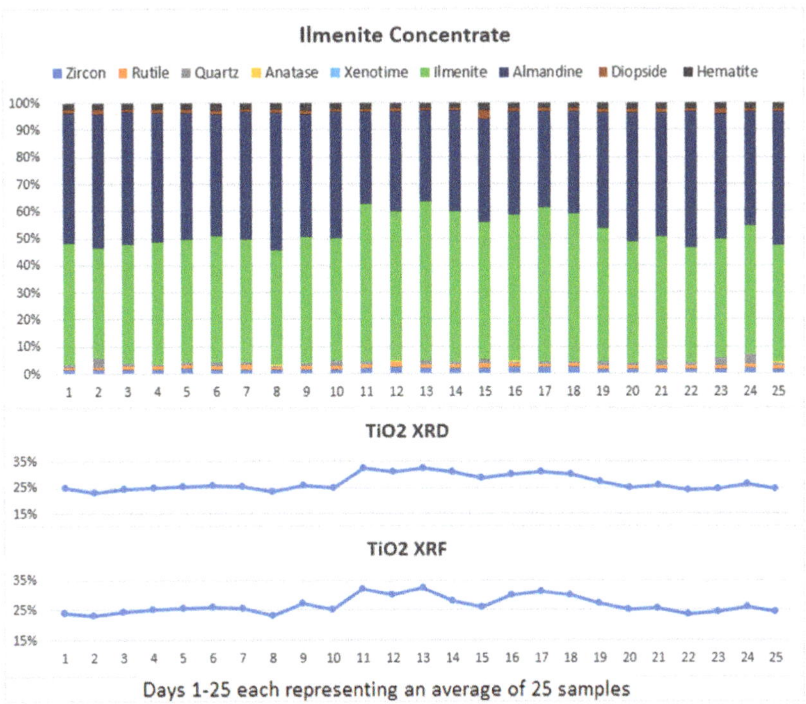

Figure 9. Variation of mineral compositions (wt%) of ilmenite concentrate, TiO_2 content (wt%) calculated from XRD results, as well as determined by XRF of 25 samples representing 25 days, (*x*-axis).

Similar scenarios apply to the garnet and zircon products. Average mineral abundances obtained over time can be used to predict and optimize shipment blends and compositions to match target values (see Figure 10 and Table 1).

Table 1. Average mineral abundancies (weight%) of 25 weekly composite samples each, and target values.

	Ilmenite Concentrate		Garnet Concentrate		Zircon Concentrate	
	Product Average	**Target Value**	**Product Average**	**Target Value**	**Product Average**	**Target Value**
Zircon	2	0–4	1		76	>67
Rutile	1	0–3	1		17	>15
Quartz	1		4		3	
Anatase	0		0		1	
Clinopyroxene	1		13		0	
Ilmenite	46	35–55	2		1	
Hematite	2		1		1	
Garnet	46	45–54	72	70–75	3	
Plagioclase	0		6		0	

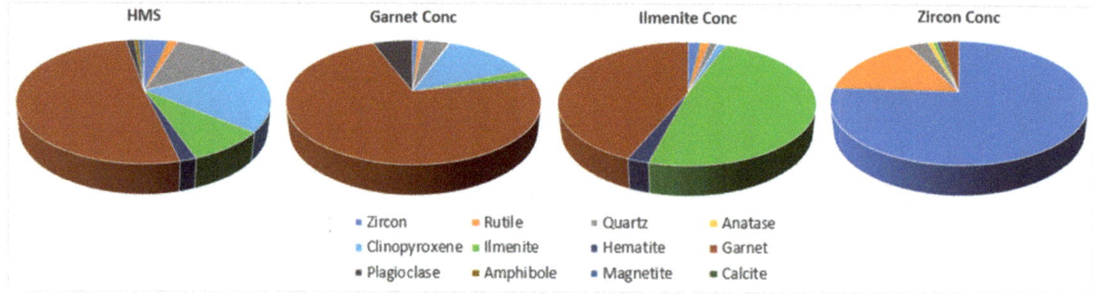

Figure 10. Average mineral abundancies of HMC and product streams of 25 weekly composite samples each.

3.3. Case Study III: Process Control during Ilmenite Smelting

Ilmenite smelting is one of the methods to upgrade the iron-titanium oxide mineral ilmenite ($FeTiO_3$) to a high-titanium feedstock for rutile pigment manufacture. In ilmenite smelting, the FeO content of the feed material ($FeO \cdot TiO_2$) is decreased by reduction with carbon. In a parallel reaction, a significant amount of TiO_2 is reduced to Ti_2O_3. The amount of Ti_2O_3 is an important parameter to control the smelting process to produce an acceptable slag product, [11–13]. Improving the understanding of a slag product, and the factors relevant in the production of a liquid slag leads to more efficiency in the smelting process.

The XRD pattern shown in Figure 11 represents a pattern of a reduced ilmenite sample. Ilmenite was reduced to predominantly rutile (TiO_2) and metallic iron. Besides metallic Fe (peak at approximately 52.5°2θ) most phases in the reduced ilmenite are titanium-based.

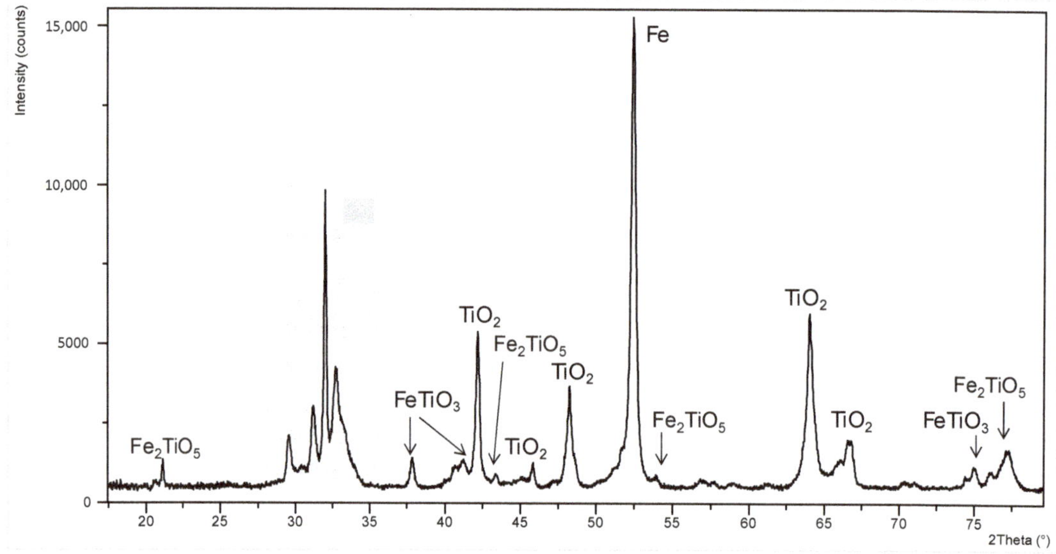

Figure 11. Phase identification for a reduced ilmenite sample, gray = detailed view Figure 12.

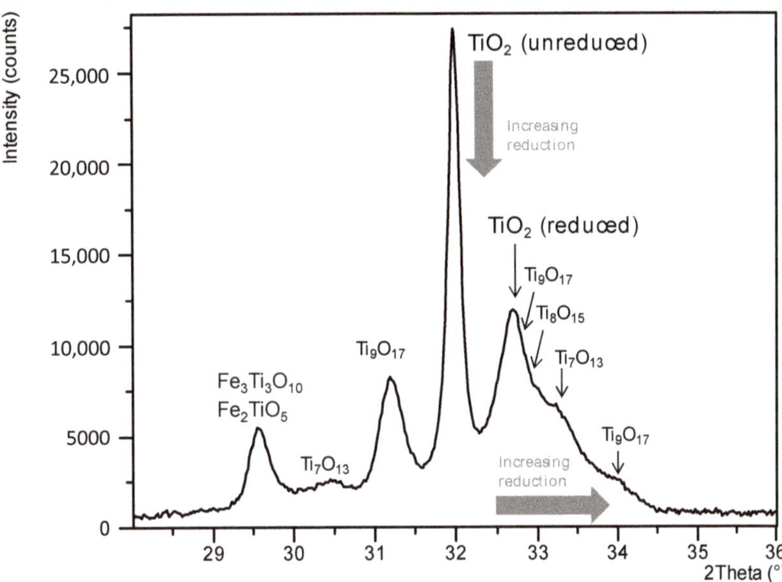

Figure 12. Detailed measurement showing different titanium oxides of a reduced ilmenite sample.

A significant variation in the different titanium phases was observed in the range $29° \leq 2\theta \leq 35°$. In the zoomed view in Figure 12 series of satellite reflections, so-called Magnéli phase reflections [14] are detected around the main unreduced rutile peak. These subpatterns are partly superposed. Magnéli phases are a suite of super-lattices derived from the rutile structure due to systematic oxygen vacancies. They have stoichiometries of Ti$_n$O$_{2n-1}$ where n $\geq$ 4. In the limit of large n, the stoichiometry approaches that of rutile. Magnéli phases form in high temperatures under reducing kiln conditions, which are required to produce synthetic rutile of higher grade (i.e., higher titanium content) at commercially reasonable throughputs [15]. In a well-run kiln, the unreduced rutile peak at 32.0°2θ is less intense than the satellite peaks from partially reduced rutile. The peak at approximately 32.6°2θ is caused by reduced rutile. As the degree of reduction increases, the peak shift to higher 2θ values and forms doublets with highly reduced material. Therefore, the degree of reduction during the smelting process can be estimated by:

(a)   Intensity ratio between the unreduced rutile peak and the reduced rutile peaks
(b)   Peak position of the reduced rutile peak and the shifts of the Magnéli peaks towards higher 2θ values.

Future work will focus on a full pattern Rietveld quantification of the different titanium oxides, especially the various Magnéli phases. This requires structural data of all phases present. A quantification of the different phases will enable even more efficiency during Ti slag production.

4. Conclusions

All three case studies demonstrated that XRD is a fast, accurate and flexible method to monitor heavy mineral mining and processing. Cluster analysis can be used to monitor and control the performance of spirals for heavy mineral concentration. The quantification of the heavy minerals during separation ensures optimal product quality and efficient stockpiling and processing. Knowledge about the degree of reduction during ilmenite smelting ensures optimal kiln operation and energy consumption. Monitoring of the different titanium oxide phases can help to produce high-quality titanium slag.

Insights into the mineralogical composition of material streams during heavy mineral processing allows fast counteractions on changing raw materials and process conditions during concentration, separation and smelting.

Author Contributions: Conceptualization, U.K. and S.M.C.V.; methodology, U.K. and S.M.C.V.; validation, U.K. and S.M.C.V.; formal analysis, U.K. and S.M.C.V.; investigation, U.K. and S.M.C.V.; resources, U.K. and S.M.C.V.; data curation, U.K. and S.M.C.V.; writing—original draft preparation, U.K. and S.M.C.V.; writing—review and editing, U.K. and S.M.C.V.; visualization, U.K. and S.M.C.V.; supervision, U.K. and S.M.C.V.; project administration, U.K. and S.M.C.V. All authors have read and agreed to the published version of the manuscript.

Funding: This research received no external funding.

Data Availability Statement: All data were gathered and treated at XRD Analytical and Consulting in Pretoria (South Africa) and Malvern Panalytical B.V laboratories in Almelo (The Netherlands).

Conflicts of Interest: The authors declare no conflict of interest.

References

1. König, U.; Spicer, E. X-ray Diffraction (XRD) as fast industrial analysis method for heavy mineral sands in process control and automation–Rietveld refinement and data clustering. In Proceedings of the 6th International Heavy Minerals Conference, Zulu Hluhluwe, South Africa, 10–14 September 2007; pp. 189–195.
2. Spicer, S.M.C.; Verryn, S.; Deysel, K. Analysis of Heavy Mineral Sands by Quantitative X-Ray Powder Diffraction and Mineral Liberation Analyser—Implications for Process Control. *ICAM Aust.* **2008**, *1*, 8–10.
3. König, U. Applications of X-ray diffraction in heavy mineral processing–Potential implications for concentration and ilmenite reduction. In Proceedings of the 9th International Heavy Minerals Conference, Visakhapatnam, India, 27–29 November 2013; paper 1315, Volume 1, 10p.
4. Available online: http://www.crystallography.net/cod/ (accessed on 13 August 2021).
5. Degen, T.; Sadki, M.; Bron, E.; König, U.; Nénert, G. The HighScore suite. *Powder Diffr.* **2014**, *29* (Suppl. S2), S13–S18. [CrossRef]
6. Rietveld, H.M. A profile refinement method for nuclear and magnetic structures. *J. Appl. Cryst.* **1969**, *2*, 65–71. [CrossRef]
7. Rietveld, H.M. Line profiles of neutron powder diffraction peaks for structure refinement. *Acta Crystallogr.* **1967**, *22*, 151–152. [CrossRef]
8. Young, R.A. *The Rietveld Method*; International Union of Crystallography; Oxford University Press: Oxford, UK, 1993; 298p.
9. Kelley, L.A.; Gardner, S.P.; Sutcliffe, M.J. An automated approach for clustering an ensemble of NMR-derived protein structures into conformational-related subfamilies. *Protein Eng.* **1996**, *9*, 1063–1065. [CrossRef] [PubMed]
10. Liao, B.; Chen, J. The application of cluster analysis in X-ray diffraction phase analysis. *J.Appl. Cryst.* **1992**, *25*, 336–339. [CrossRef]
11. Zietsman, J.H.; Pistorious, P.C. Process mechanism in ilmenite smelting. *J. South Afr. Inst. Min. Metall.* **2004**, *105*, 653–660.
12. Pistorius, P.C. Ilmenite smelting: The basics. *J. South Afr. Inst. Min. Metall.* **2008**, *108*, 35–43.
13. Gupta, A.K.; Aula, M.; Pihlasalo, J.; Mäkelä, P.; Huttula, M.; Fabritius, T. Preparation of synthetic titania slag relevant to the industrial smelting process using an introduction furnace. *Appl. Sci.* **2021**, *11*, 1153. [CrossRef]
14. Anderson, S.; Collen, B.; Kruuse, G.; Kuylenstierna, U.; Magneli, A.; Pestmalis, H.; Asbrink, S. Identification of titanium oxides by X-ray powder patterns. *Acta Chem. Scand.* **1957**, *11*, 1653. [CrossRef]
15. Pax, R.A.; Steward, G.A. Quantitative determination of phases using Mossbauer spectroscopy and X-ray diffraction: A case study of the Fe-Ti-O system. In Proceedings of the 36th Annual Condensed Matter and Materials Meeting 2012, Wagga Wagga, NSW, Australia, 31 January–3 February 2012.

Article

Characterization and Suitability of Nigerian Barites for Different Industrial Applications

Itohan Otoijamun [1,2,*], Moses Kigozi [1,3], Adelana Rasak Adetunji [4] and Peter Azikiwe Onwualu [1]

[1] Department of Materials Science and Engineering, African University of Science and Technology, Abuja P.M.B 681, Nigeria; mkigozi@aust.edu.ng (M.K.); aonwualu@aust.edu.ng (P.A.O.)
[2] National Agency for Science and Engineering Infrastructure, Garki Abuja P.M.B 391, Nigeria
[3] Department of Chemistry, Faculty of Science and Education, Busitema University Uganda, Tororo P.O. Box 236, Uganda
[4] Materials Science and Engineering Department, Kwara State University, Malete Kwara P.M.B 1530, Nigeria; aderade2004@yahoo.com
* Correspondence: oitohan@aust.edu.ng; Tel.: +234-80-3849-0365

Abstract: This work aimed to characterize barite samples from selected different locations in Nigeria and determine their suitability for various industrial applications. The properties determined include mineralogy, chemical composition, morphology, functional groups, and specific gravity. Samples were obtained from ten locations in Nasarawa and Taraba states as well as a standard working sample (WS) obtained from a drilling site. The samples were characterized using scanning electron microscope and energy dispersive X-ray (SEM-EDX), Fourier infrared analysis (FTIR), and X-ray diffraction (XRD). Specific gravity (SG) was determined using the pycnometer method. Results of SEM-EDX analysis show that the WS has a Ba-S-O empirical composition of 66.5% whereas these of the ten samples investigated vary between 59.36% and 98.86%. The FTIR analysis shows that the functional groups of S-O, SO_4^{2-}, Ba-S-O, OH of the ten samples match that of the WS. Results of XRD show that the ten samples have the same mineralogical composition as the WS and all meet American Petroleum Institute (API) standards for industrial barite. Similar matching results are shown from EDXRF spectra intensity, position, and composition analysis of the ten samples compared to the WS. Specific gravity (SG) results show that six out of the ten samples have SG above 4.2 which is the recommended minimum for the American Petroleum Institute (API) standard. The other four samples will require beneficiation to meet the standard for drilling mud application. Using all the parameters of the assessment together, results show that while some (6) of the samples can be used for drilling fluid application, some (4) require beneficiation but all ten samples can be used for other industrial applications including healthcare, construction, plastic, cosmetics, paper, and rubber industries. The results of the study can be used for value addition in developing beneficiation procedures, processes, and technology for purification along with new materials for the industries.

Keywords: barite; mineralogy; industrial application; beneficiation; specific gravity

Citation: Otoijamun, I.; Kigozi, M.; Adetunji, A.R.; Onwualu, P.A. Characterization and Suitability of Nigerian Barites for Different Industrial Applications. *Minerals* **2021**, *11*, 360. https://doi.org/10.3390/min11040360

Academic Editors: Herbert Pöllmann and Uwe König

Received: 20 February 2021
Accepted: 24 March 2021
Published: 30 March 2021

Publisher's Note: MDPI stays neutral with regard to jurisdictional claims in published maps and institutional affiliations.

1. Introduction

The Federal Government of Nigeria is currently implementing the National Economic Recovery and Growth Plan (ERGP) aimed at re-directing the economy back to the path of recovery [1]. A major aspect of the plan is diversification of the economy away from oil and increasing the local content in operations of the oil industry. One way of diversifying the economy is by developing the mining sector, including adding value to extracted minerals. Approximately 85% of barite goes into the oil industry, about 10% into the chemical industry, 5% into the filler market. Barite is used as a weight density agent in drilling mud for gas and oil exploration to avoid the high-pressure formation and prevent blowouts. This is compressing the high pressure created by the drill bit as it passes through various formations with different characteristics. The deeper the drilling hole, the more barite is required for the total mud mix. For oil drilling, a specific gravity of barite is the only property checked. Other chemical and physical properties are needed for other barite applications. The physical appearance of barite used in drilling petroleum wells can be black, blue, brown, or grey dependent on the ore body. The used barite must be dense enough so that its specific gravity is greater than 4.1 and smooth not to damage the drill bit [2,3]. One of the minerals being promoted by the Federal Ministry of Mines and Steel Development is barite [4]. This mineral is not only useful in the oil industry but can also be used for other industrial applications. Most of the current use of barite in the oil industry is imported [4]. Therefore, there is a need to promote the exploitation and use of barite for the oil industry and other industrial applications.

One way of diversifying the economy is by developing the mining sector, including adding value to extracted minerals. This is mainly in the oil industry which depends on the appreciable supply of barite ore as a key constituent of drilling mud used to stabilize the oil well; prevent blow-outs; remove drill cuts by the fluid. It is a key constituent of drilling mud, which is the fluid pumped into the oil or gas well to lubricate the bit and drill stem, removes rock chips, prevents a collapse of well walls, and prevents blowouts if over pressured strata are encountered [5–8]. Barite is chemically stable, making it useful as an additive in the manufacturing of different products like rubber, paints, enamels, plastics, paper goods, wallpapers, asbestos goods, glass, and ceramics. Moreover, it is used in radiology for X-rays of the intestines and to make high-density concrete resistant to nuclear radiation [7]. Barite ($BaSO_4$) is crucial to the oil and gas industrial application. This is due to a key constituent of the drilling mud used in oil and gas wells. Additionally, elemental barium is an additive in ceramic glazes, optical glass, paint, and other products. Barite deposits are categorized into different main types which include; bedded-volcanic, bedded-sedimentary, vein, cavity-fill, and metasomatic and residual. Bedded-sedimentary deposits are found in sedimentary rocks with properties of high biological productivity during sediment accumulation and they are the major sources of barite production that account for the majority of barite reserves worldwide [8]. In recent years, barite has found usage in brake shoe linings, noise reduction in engine compartments, and spark-plug alloys [9].

Barite is a heavy mineral that normally occurs with Pb-Zn ore, barite vein, barite-fluorite vein deposit, strata bound SEDEX-type deposit among other deposits as a gangue mineral, in sedimentary deposits, and rarely in salts [8]. It is usually mined as barium content. It occurs either in crystalline form, as tabular, prismatic, or bladed crystals. The pure crystals are often colorless, cream-colored, or white, but may also acquire various colors based on the impurities it contains. Granitic rocks characteristically have a somewhat higher content of barium than average continental crust, and basaltic rocks characteristically have lower barium content. The range of barium content of shales spans approximately the same range as the barium content of granitic rocks [8]. Some smaller mines exploit barite in veins, which formed when barium sulfate was precipitated from hot subterranean waters. In some cases, barite is a by-product of mining lead, zinc, silver, or other metal ores [10], in the paper and rubber industries, as a filler or extender in cloth, ink, and plastics products; in radiography ("barium milkshake"); as getter (scavenger) alloys in vacuum tubes; deoxidizer for copper; lubricant for anode rotors in X-ray tubes, spark-plug alloys, and white pigment. Other uses of barite include as an additive for friction materials, rubbers, plastics, paints, feedstock for chemical manufacturing, and shielding in X-ray and gamma-ray applications [8,11].

The status of barite mining activities in Nigeria currently shows that the barite quality from these different localities proves that Nigeria does not necessarily need to import high grade or any other specification of barite from foreign countries for its usage in the desired industries [12]. However, a large percentage of the barite used in the oil industry is imported. In 2020 alone, this is valued at about $96 million, and the estimated consumption is 440,000 metric tons for this year. To boost the mineral sector, the Federal Government of Nigeria has taken initiatives to encourage local mining, beneficiation, and mining of barite for the industries [4]. The valuation activities are usually done by the Federal Ministry of Mines and Steel Development (MMSD), Nigerian Geological Survey Agency (NGSA), and Nigerian Content Development and Monitoring Board (NCDMB).

Drilling activity accounts for nearly 95 percent of domestic consumption and about 90 percent of global consumption. Economic deposits of barite are relatively common and are found in many countries [8]. Literature shows that some Nigerian barites are suitable for drilling fluid formulations [13]. In another study, it was shown that barite and oil drilling fluid additives affect reservoir rock characteristics [13]. Another study reported on the effect of barite and ilmenite mixture on enhancing the drilling mud weight [14]. Mohamed et al. showed that barite sag can be prevented in oil-based drilling fluids using a mixture of barite and ilmenite as weighting material [7]. The future of drilling-grade barite weight material was presented by [15]. There are frequently auxiliary criteria used to compare deposits of barite. Some raw materials have sensitive costs which include transportation of raw materials and market, involving land and sea costs. This can inform how much surplus remains for mining and milling for the deposits to be considered economically viable. The grade of the deposits also renders economic viability including the revenue for the cost of mining and milling [16]. The mining process may have little or no effect on the policy change of the companies or management. For example, the change of management from a GmbH and Co. KG to a GmbH did not affect the economic identity of the legal entity and never lead to a transfer of assets. The alteration was limited to changing the legal control while preserving the legal identity. This is because there is no act of company asset transfer and results in no exchange of services, hence no legal taxes charged [17]. Characterization results from previous work were mostly centered on discovering the chemical composition (XRF), mineralogy (SEM), organic, polymeric (FTIR), specific gravity (SG), needed for the same oil industry. The use in other industries is not well documented even though it plays an important role in these industries.

This work aims at the characterization of barite samples from different locations in Nigeria and the determination of their suitability for different industrial applications. The properties determined include mineralogy, chemical composition, morphology, functional groups, specific gravity, and the physical appearance of the powdered samples.

2. Materials and Methods

2.1. Study Location

Barite occurs in various locations within the Benue Trough. These include Adamawa, Benue, Cross River, Ebonyi, Gombe, Nasarawa, Plateau, and Taraba. But studies have shown that the major producing states are Benue, Cross River, Nasarawa, and Taraba. Samples used in the study were taken from Nasarawa (NS) and Taraba states (TS). The samples from Nassarawa state were obtained from nine different locations: Azara vein 1, Azara vein 17, Azara vein 18, Aloshi, Keana, Kumar, Ribi, Sauni, and Wuse. One sample was from Ibi in Taraba state. A sample was picked from one active well as a working sample was obtained from Port Harcourt (Rivers state) from a drilling site. These state locations are shown on the map of Nigeria in Figure 1.

2.2. Sample Collection

Different barite deposit sites were visited, including Nasarawa and Taraba states of Nigeria with representative samples collected from mining pits being worked by artisanal miners. Rocks exposure within the deposits were studied to understand the lithology of the deposits. Samples were collected across the veins and stored in sample bags with name tags as highlighted in the supplementary information.

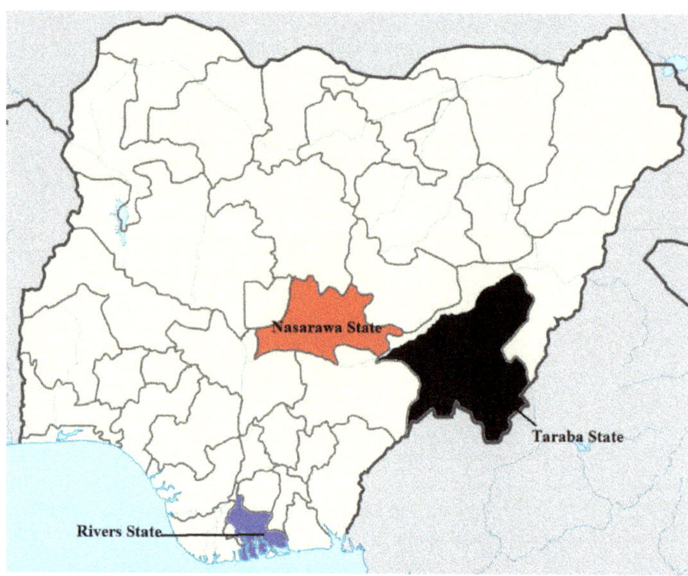

Figure 1. Map of Nigeria showing active mining sites for various minerals [18]. The location and local Government area of the mine sites is shown in Table S1 (Supplementary Materials).

2.3. Sample Characterization

The samples were studied in the field as hand specimens and further characterized for physical and chemical properties which include: scanning electron microscope and energy dispersive X-ray (SEM-EDX, Carl ZEISS, Bangalore, India) for the in-depth surface analysis, the FTIR (Fourier transform infrared, Bruker Optik GmbH Vertex 70, Ettlingen, Germany), the XRD (X-ray diffraction, Riguku Smartlab Autosampler (RIGAKU Corp., Tokyo, Japan), the EDXRF (Malvern Panalytical B.V., Almole, Netherland) for oxides analysis [18–20]. SG was used to evaluate the physical, mineralogical, and chemical properties of the barite ores from the various locations [21–23] as described in the supplementary information.

Sample from Aloshi Nasarawa state was labeled as (NT) and Sample from Ibi, Taraba state was labeled as (TS); all other samples were labeled by the name of the location it was obtained from. The sample which was picked from the Port Harcourt drilling site labeled as (WS) was taken as the working standard.

3. Results and Discussion

The scanning electron microscope (SEM) and energy dispersive X-ray (EDX) analysis of morphology and element percentage composition data are shown in Figure 2 and supplementary information Figure S1 (Supplementary Materials). The morphology of samples NS, TS, Azara vein 1, Azara vein 17, Azara vein 18, Keana, Kumar, Ribi, Sauni, Wuse, and WS, showed surfaces of barite embedded form of material structure with clear unit boundaries. The working standard sample (WS) showed a fine surface in the morphology monograph. The surface chemical analysis was carried out with EDX, revealing the different elemental compositions of the material in-depth surfaces (Table 1). Sample NS revealed a composition of Ba as 46.73%, S as 33.55%, and oxygen as 1.93%. This gave 82.21% of the Ba-S-O empirical composition. Sample TS showed a composition of Ba as 50.37%, S as 35.63%, and oxygen as 1.87%. This gave 87.87% of the Ba-S-O empirical composition. Sample Azara vein 1 showed a composition of Ba as 51.73%, S as 43.36%, and oxygen as 2.11%. This gave 97.2% of the Ba-S-O empirical composition. Sample Azara vein 17 showed a composition of Ba as 38.59%, S as 36.84%, and oxygen as 2.88%. This gave 78.31% of the Ba-S-O empirical composition. Sample Azara vein 18 showed a composition of Ba as 34.79%, S as 31.48%, and oxygen as 3.74%. This gave 70.01% of the Ba-S-O empirical composition. Sample Keana showed a composition of Ba as 50.12%, S as 42.75%, and oxygen as 3.14%. This gave 96.01% of the Ba-S-O empirical composition. Sample Kumar showed a composition of Ba as 31.36%, S as 23.82%, and oxygen as 4.18%. This gave 59.36% of the Ba-S-O empirical composition. Sample Ribi showed a composition of Ba as 51.38%, S as 44.17%, and oxygen as 2.59%. This gave 98.14% of the Ba-S-O empirical composition. Sample Suani showed a composition of Ba as 51.17%, S as 44.75%, and oxygen as 3.94%. This gave 99.86% of the Ba-S-O empirical composition. Sample Wuse showed a composition of Ba as 51.42%, S as 42.58%, and oxygen as 2.98%. This gave 96.98% of the Ba-S-O empirical composition. Sample WS showed a composition of Ba as 38.26%, S as 25.89%, and oxygen as 2.35%. This gave 66.5% of the Ba-S-O empirical composition. The working standard sample showed a much lower element composition empirical percentage indicating that most of the samples will be suitable for drilling purposes [23–26].

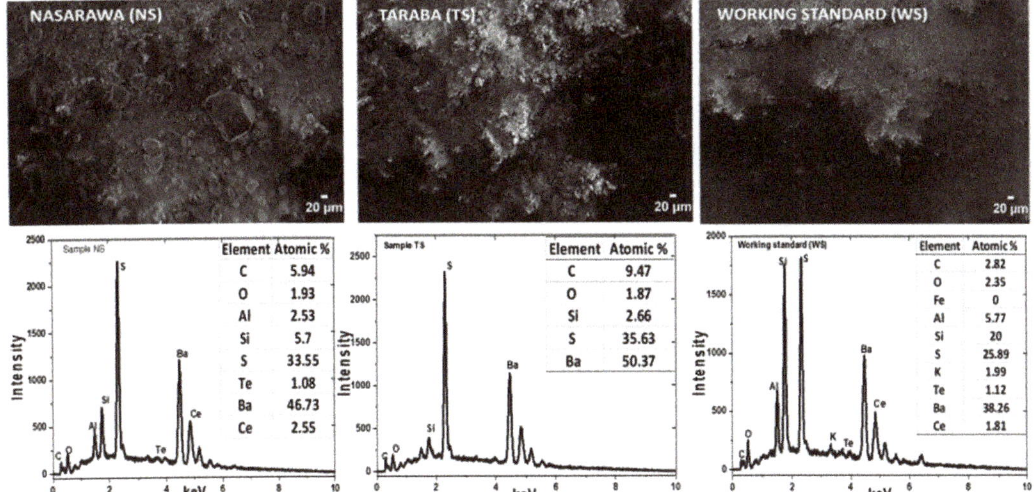

Figure 2. SEM-EDX morphology and sample element atomic percentages.

Table 1. Energy-dispersive X-ray spectroscopy (EDS) atomic percentage of elements from barite mineral samples from different mining sites.

Samples	EDS Elemental Percentage Composition										
	Ba	S	O	Fe	Al	Si	Te	Ce	K	La	Nb
NS	46.73	33.55	1.93		2.53	5.7	1.08	2.55			
TS	50.37	35.63	1.87			2.66					
Azara Vein 1	51.73	43.36	2.11			2.64	0.17				
Azara Vein 17	38.59	36.84	2.88			21.66		0.03			
Azara Vein 18	34.79	31.48	3.74	23.89	2.2	3.25	0.29			0.36	
Keana	50.12	42.75	3.14			2.21		0.02			
Kumar	31.36	23.82	4.18	28.64		11.68	0.32				
Ribi	51.38	44.17	2.59				0.07				1.79
Sauni	51.17	44.75	3.94				0.14				
Wuse	51.42	42.58	2.98			2.05					
WS	38.26	25.89	2.35		5.77	20	1.12	1.81	1.99		

The FTIR technique was used for bond identification for chemical structures in the materials. This is used as a fingerprint for the mineral group identification and information about the structure of the mineral. The FTIR spectra for the samples are shown in

Figure 3a,b and supplementary information Figure S2. All samples show several peaks in both fingerprint and functional regions which are comparable to the working standard material (WS). There are strong peaks at 600, 1059, 1197 cm^{-1} wavenumber in the fingerprint region and 1640, 2047, and 3468 cm^{-1} wavenumber in the functional region. The peaks at 600 cm^{-1} with strong transmittance indicate the Ba-S-O polyhedral stretching which is the sheet structure of barite minerals. The peak at 1059 cm^{-1} demonstrates the triple asymmetric S-O stretching in barite which also indicates the stretching of SO$_4{}^{2-}$ tetrahedral. The peak at 1197 cm^{-1} shows the asymmetric and bond vibration [27]. In the functional region, the samples depicted peaks at different wavenumbers. The peak at 1640 cm^{-1}, which is medium, indicates the stretching vibration of the oxygen group. This is an indication of the S-O bond for the structure. The peak at 2047 cm^{-1} shows the formation of the Ba-S-O bond stretching vibration in the functional region. This contributes to the empirical structure of barite [28]. The peak formation at 3468 cm^{-1} indicates the OH stretch which is due to the formation of crystalline structure in the material [27,29]. The peaks in the samples NS and TS are matching with those of the working standard sample. This means that the samples have the same functional groups which match the working standard sample. The sample results are identical to others in the literature [30].

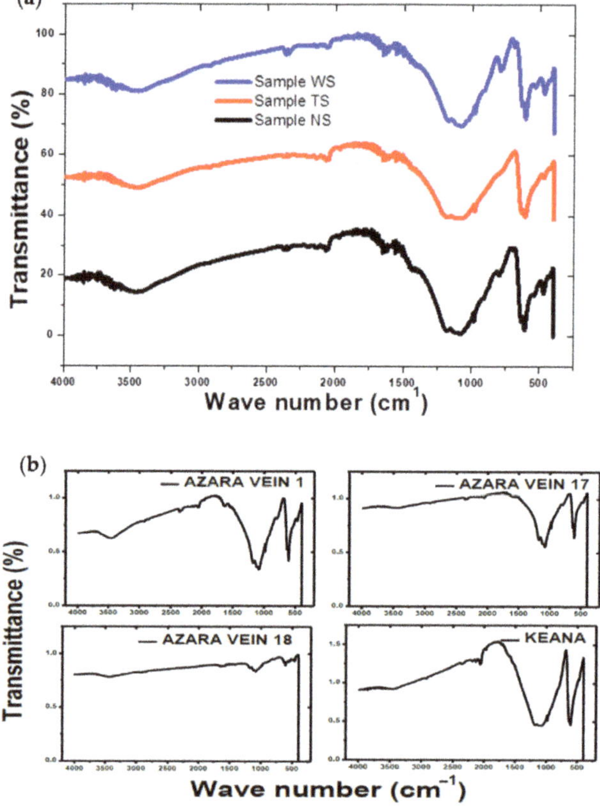

Figure 3. (**a**) FTIR spectra for the barite samples from different mining sites; (**b**) FTIR spectra for the barite samples from different mining sites.

The XRD spectra for all the samples are shown in Figure 4. Samples exhibit the main peak at $2\theta = 28.75°$ with d-spacing of 3.102 Å with a plane of (211). Other peaks appeared at $2\theta = 26.85°$ with d-spacing of 3.32 Å and plane of (102), at $2\theta = 25.8°$ and d-spacing of 3.44 Å and plane of (210), and $2\theta = 42.89°$ with d-spacing of 2.10 Å and plane of (112). The XRD data revealed a structure formation with a phase of $BaSO_4$ in the chemical form of $Ba_4S_4O_{16}$ and the calculated density of 4.47 g/cm^3. Sample NS exhibited different peaks at slightly no difference in position from other samples with a peak from $2\theta = 25.86°$ with d-space of 3.44 Å and plane of (210), the main peak at $2\theta = 28.75°$ with d-spacing of 3.10 Å and plane of (211) same as other samples. Other peaks appeared at $2\theta = 26.85°$ with d-spacing of 3.32 Å and plane of (102) also at $2\theta = 31.54°$ with d-spacing of 2.83 Å with a plane of (112) for all the samples. The samples also revealed a structure formation with a phase of $BaSO_4$ and the same chemical form of $Ba_4S_4O_{16}$ and the calculated density of 4.47 g/cm^3. XRD peaks of the crude Barite powder, which indicates peaks corresponding to Barite with chemical formula $BaSO_4$ on ICPDS card number 00-024-0020 [16,29,30].

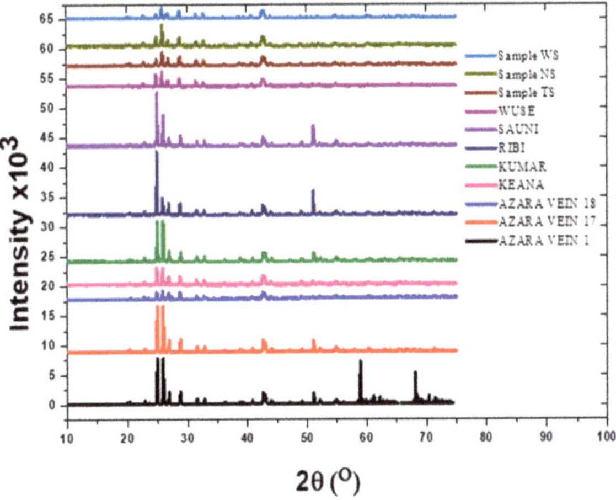

Figure 4. XRD spectra for samples from different mining sites.

The working sample (WS) depicted the main peaks at $2\theta = 28.737°$ with d-spacing of 3.104 Å and a plane of (211) as the main peak. Other peaks appeared at $2\theta = 25.841°$ with d-spacing of 3.445 Å and plane of (211), $2\theta = 26.837$ with d-spacing of 3.319 Å and plane of (102), $2\theta = 31.522°$ with d-spacing of 2.836 Å and plane of (112), $2\theta = 32.797°$ d-spacing of 2.729 Å and plane of (020), $2\theta = 42.871°$ with d-spacing of 2.11 Å. The main peaks in the working sample at 2θ of 28.737, 25.841, 26.837, 32.797, and 42.871° match with the main peaks in the samples. This means that the samples have the same composition which was also shown by FTIR results (Figure 3a,b). The amount of barium sulphate shown by the chemical analysis revealed that the tested local barite samples are following the API requirements for barite [12,30].

The samples were further analyzed by EDXRF as shown in the supplementary information (Figure S3–S10 in Supplementary Materials), show the spectra of a K feldspar and their quantification results as listed in Table 2. All the samples match well the chemical data of the working sample (WS) as shown in other figures in the supplementary information (Figure S3–S10). In general, the mineral distribution maps of both classifications of the samples correspond well with that of the working sample in the spectra intensity, position, and composition. Texture and grain structures of samples' complex intergrowth are noticeably well in the monographs. A few variances can be found in details of the other trace elements which were able to be detected in the microstructures such as micro-perthitic intergrowth in sample WS and other samples. This was due to the smaller beam diameter and the crushing limitations to a few micrometers.

Table 2. EDXRF elemental oxide percentage composition of the samples from different mining sites.

OXIDES/ Elements	Percentage of Elemental Oxide Composition for Samples										
	WS	NS	TS	AZARA VEIN 1	AZARA VEIN 17	AZARA VEIN 18	KEANA	KUMAR	SAUNI	RIBI	WUSE
Fe_2O_3	0.175	0.307	0.028	0.101	0.292	9.76	0.028	0.553	0.165	0.102	0.015
SiO_2	1.831	1.878	1.164	6.752	5.036	10.11	1.709	2.989	0.592	4.355	1.362
Al_2O_3	0.722	0.681	0.667	1.193	0.914	2.415	0.783	1.281	0.397	1.229	0.463
MgO	1.731	0.34	1.14	0.18	0.19	0.05	1.69	0.53	0.28		1.18
P_2O_5	0.023	0.05		0.023	0.021	0.056	0.033			0.05	
SO_3	17.3	7.609	7.687	13.73	16.3	10.81	16.65	16.53	12.3	16.29	13.44
TiO_2		6.353	6.572								
MnO					0.013	0.421		0.014	0.069		
CaO		0.389	0.166	0.01	0.004	0.113				0.002	
K_2O	0.030	0105	0.092	0.092	0.048	0.496	0.029	0.055	0.015	0.116	0.008
CuO	0.002	0.001	0.001	0.002	0.002	0.015	0.002	0.003	0.005	0.001	0.003
ZnO		0.001				0.045					
Cr_2O_3	0.023			0.03	0.028	0.002	0.032	0.018	0.011	0.034	0.02
PbO	0.281	0.006	0.001		0.001		0.274	0.001	0.001		
Rb_2O	0.002			0.001		0.001	0.002				
Cl	0.451	0.494	0.644	0.432	0.526	0.469	0.447	0.556	0.314	0.515	0.322
BaO	30.45	18.04	18.62	26.32	31.42	24.18	32.57	31.32	22.94	31.64	24.33
Ta_2O_5		0.017	0.003		0.003	0.001				0.002	
WO_3	0.312				0.016	0.106	0.304				
SrO	2.855	0.546	4.687	4.258	3.309	2.72	2.834	5.68	2.666	1.421	6.148
CeO_2	1.712			1.42	1.744	1.513	1.661	1.772	1.3	1.658	1.234
ThO_3				0.001	0.001	0		0.001	0.001		
Y_2O_3				0.002	0.017	0.002					
Nb_2O_5	0.001	0.001	0.001	0.002	0.002	0.002	0.002	0.002		0.002	0.02
SnO_2								0.017			
Sb_2O_3	0.001		0.11		0.001	0.001	0.001			0.001	
Cs		0.156	0.055								

The specific gravity of the samples was determined and the results are shown in Table 3 for all the samples from different mining sites. In reference to the American Petroleum Institute (API) standard specification of not less than 4.15, barite is used to increase the apparent density of a liquid drilling fluid system. Most of the samples showed higher SG apart sample TS, NS, and Azara vein 18 with lower SG than the API standard. This makes barite [$BaSO_4$] the most common weighting agent used today. It is a mined material ground to an API specification such that particle sizes are predominantly in the 3 to 74 μm. The results in Table 3 displayed that the specific gravity (SG) of the samples is higher than that of the working sample. This implies that these samples from the field can be used as a replacement for the working sample.

Table 3. The specific gravity of samples from different mining sites.

Sample	Specific Gravity (g/cm^3)
SAMPLE TS	4.0087
SAMPLE NS	3.8122
AZARA VEIN 1	4.2138
AZARA VEIN 17	4.3761
AZARA VEIN 18	4.0106
KEANA	4.4052
KUMAR	4.4000
RIBI	4.4200
SAUNI	4.3800
WUSE	4.4000
WORKING SAMPLE (WS)	3.6001

4. Suitability of Barites for Industrial Applications

As noted earlier, barites can be used in several industries including oil and gas (drilling fluid formulation); healthcare (X-ray, Plaster of Paris, making barium solution for stomach and intestine reflections); construction (paints, blocking emission of gamma-rays through walls in hospitals, power plants, and laboratories); plastic (filler), cosmetics, paper (filler), and rubber industries. The suitability of the barite samples for different industrial applications was evaluated. The samples which were analyzed from selected mining sites have a barite percentage ranging between 30 to 50% of barium content before beneficiation and removal of other impurities. This means that the barite from those different mining sites can be used for preliminary mining activities and other application which require a barite composition in the range of 18 to 34% like kilns, mud drilling, construction, among others [29–31]. If the materials are to be used for the main applications, some of the impurities need to be removed which will help to increase the composition percentage of barite. The removal of silica from the samples can increase the purity of the material for applications where silica is not needed [6]. Some impurities are used with barite to improve the properties of the materials as shown in Table 4. All samples can be beneficiated and hence improved for different applications as shown in Table 4 because they have the required chemical constituents. When the unwanted impurities are removed, the percentage composition of barite can be increased to suit the required application. In chemical manufacturing, some samples were eliminated because they lack $CaCO_3$ in their initial composition.

Table 4. American Petroleum Institute (API) and American Society for Testing and Materials (ASTM) general specification standards for various uses of barite ores for different industrial applications.

Barite Application	(%) $BaSO_4$ Sinimum std	Constituents	Specific Gravity Minimum std (g/cm^3)	Study Samples Suitable for Application after Purification
Oil well drilling	90		4.15	All samples apart from TS
Chemical manufacturing	97	SiO_2, $CaCO_3$, Al, Fe	4.0	NS, TS, Azara 1, Azara 17, Azara 18, Ribi
Paint manufacturing	95		4.45	All samples
Glass	90–96	SiO_2, Al, Fe		All samples
Pharmaceuticals	97	Fe_2O_3, SiO_2, Al_2O_3		All samples
Rubber	99.5	SiO_2		All samples
Asbestos products	90	Fe_2O_3, SiO_2, Al_2O_3		All samples
Plastering	95	SiO_2, Al_2O_3		All samples
Cement	95	SiO_2		All samples

5. Conclusions

The ten barite samples were obtained from different mining locations in the Nasarawa and Taraba states of Nigeria. Their properties were determined and compared with a standard working sample used by an oil industry operator in Nigeria. Using different characterization parameters (SEM-EDX, FTIR, XRD, SG, and physical appearance) exhibiting the molecular structure of $BaSO_4$. The characterization has shown that some (6) of the samples can be used for drilling fluid formulation for the oil and gas industry due to their good specific gravity greater than 4.15 for API. Samples like TS, NS, and Azara vein require beneficiation to reach the standard for oil application due to their low specific gravity. All ten samples can be used for other industrial applications including healthcare, construction, plastic, cosmetics, paper, and rubber industries due to their level of barium content in the range of 30 to 50%. The results of the study are being used to develop beneficiation procedures, actions, and technology along with new materials for industrial applications. Different samples exhibited different colour appearances from white to off-white which be used as filler materials in paint and ceramics as shown in Table S2. These samples will further be purified by the removal of some other mineral content to increase the yield of barium concentration.

Supplementary Materials: The following are available online at https://www.mdpi.com/article/10.3390/min11040360/s1, Figure S1: SEM-EDX morphology and sample element atomic percentages, Figure S2: FTIR Spectra for the barite samples from different mining sites, Figure S3: XRF spectra for barite samples from AZARA VEIN 1, Figure S4: XRF spectra for barite samples from AZARA VEIN 17, Figure S5: XRF spectra for barite samples from AZARA VEIN 18, Figure S6: XRF spectra for barite samples from KEANA, Figure S7: XRF spectra for barite samples from KUMAR, Figure S8: XRF spectra for barite samples from RIBI, Figure S9: XRF spectra for barite samples from SAUNI, Figure S10: XRF spectra for barite samples from WUSE, Table S1: The physical appearance of the samples, Table S2: Mine site location in the Local Government Area.

Author Contributions: The research work was conceptualized by I.O. and administration supervision was done by P.A.O. and A.R.A. The methodologies were acted on by M.K. and I.O. which included: collection, preparations and analysis. Data curation, I.O. and M.K.; Formal analysis, I.O.; Funding acquisition, P.A.O.; Investigation, M.K. and A.R.A.; Methodology, I.O. and M.K.; Project administration, P.A.O.; Resources, P.A.O.; Supervision, A.R.A. and P.A.O.; Writing—original draft, I.O. and M.K.; Writing—review & editing, A.R.A. and P.A.O. All authors have read and agreed to the published version of the manuscript.

Funding: This research was funded by Pan African Materials Institute (PAMI), a regional Centre of Excellence in Materials Science and Engineering in West Africa established under the World Bank (WB) African Centres of Excellence (ACE) program, and hosted by African University of Science and Technology, Abuja, Nigeria. The grant number is AUST/PAMI/2015/5415-NG.

Data Availability Statement: The data presented in this study is contained within the present article.

Acknowledgments: The authors would like to acknowledge the support from the members of the Department of Materials science and Engineering and the entire staff of African University of Science and Technology (AUST) Abuja Nigeria.

Conflicts of Interest: The authors declare no conflict of interest.

References

1. Ezekwesili, G.E.; Okogbue, C.O.; Chidozie, I.P. Structural styles and ecomic potentials of some barite deposits in the Southern Benue Trough, Nigeria. *Rom. J. Earth Sci.* **2012**, *85*, 1.
2. Onwualu, A.; Ogunwusi, A.; Olife, I.C.; Inyang, A. *Raw Materials Development for the Transformation of the Manufacturing Sector in Nigeria*; Raw Materials Research and Developmen Council: Garki, Abuja, Nigeria, 2013.
3. Onwualu, A.; Obasi, E.; Olife, I.; Inyang, A. *Unlocking the Potentials of Nigeria's Non-Oil Sector*; Raw Materials Research and Developmen Council: Garki, Abuja, Nigeria, 2013.
4. Olamilekan, A. *Overview on the Accelerated Development of the Barite Industry in Nigeria*; Raw Materials Research and Developmen Council: Garki, Abuja, Nigeria, 2020.
5. MJohnson, M.A.C.; Pictak, M.N. *Barite (Barium) in Minerals*; U.S. Geological Survey: Reston, VA, USA, 2017.
6. Labe, N.; Ogunleye, P.; Ibrahim, A.; Fajulugbe, T.; Gbadema, S.T. Review of the occurrence and structural controls of Baryte resources of Nigeria. *J. Degrad. Min. Lands Manag.* **2018**, *5*, 1207–1216. [CrossRef]
7. Muhammad, A.B.; Kazmia, K.R.; Mehmooda, R.; Ahadb, A.; Tabbassumc, A.; Akram, A. Beneficiation Study on Barite Ore of Duddar Area, District Lasbela. Balochistan Province Pakistan. *J. Sci. Ind. Res. Ser. A Phys. Sci.* **2017**, *60*, 9–22.
8. Johnson, C.A.; Piatak, N.M.; Miller, M.M. "Barite (Barium)". In *Critical Mineral Resources of the United States—Economic and Environmental Geology and Prospects for Future Supply*; U.S. Geological Survey: Reston, VA, USA, 2017.
9. Singh, R.; Banerjee, B.; Bhattacharyya, J.P.; Srivastava, K.K. Up-gradation of Barite Waste to Marketable Grade Concentrate. In Proceedings of the XXIII International Mineral Processing Congress (IMPC) Mineral Processing Division, National Metallurgical Laboratory, Chicago, IL, USA, 9–12 September 2007; pp. 2303–2307.
10. Ariffin, K.S. *Barite (BARIUM). Miner*; EBS 425; Perindustrian: Hyderabad, India, 2016; Volume 1.
11. Al-Awad, M.N.; Al-Qasabi, A.O. Characterization and Testing of Saudi Barite for Potential Use in Drilling Operations. *J. King Saud Univ. Eng. Sci.* **2001**, *13*, 287–298. [CrossRef]
12. Oden, M.I. Barite Veins in the Benue Trough: Field Characteristics, the Quality Issue and Some Tectonic Implications. *Environ. Nat. Resour. Res.* **2012**, *2*, 21. [CrossRef]
13. Duru, U.I.; Kerunwa, A.; Omeokwe, I.; Uwaezuoke, N.; Obah, B. Suitability of Some Nigerian Barites in Drilling Fluid Formulations. *Pet. Sci. Eng.* **2019**, *3*, 46. [CrossRef]
14. Abdou, M.; Al-Sabagh, A.; Ahmed, H.E.-S.; Fadl, A. Impact of barite and ilmenite mixture on enhancing the drilling mud weight. *Egypt J. Pet.* **2018**, *27*, 955–967. [CrossRef]
15. Bruton, J.R.; Bacho, J.P.; Newcaster, J. The Future of Drilling-Grade Barite Weight Material. In Proceedings of the SPE Annual Technical Conference and Exhibition, San Antonio, TX, USA, 24–27 September 2006.

16. Wellmer, F.-W.; Dalheimer, M.; Wagner, M. *Economic Evaluations in Exploration*, 2nd ed.; Springer: Berlin/Heidelberg, Germany, 2008.
17. Harneit, J. Umwandlung einer GmbH & Co. KG in eine GmbH. *Die Bus. Judgement Rule* **2019**, *1*, 1. [CrossRef]
18. Lar, U.A.; Agene, J.I.; Umar, A.I. Geophagic clay materials from Nigeria: A potential source of heavy metals and human health implications in mostly women and children who practice it. *Environ. Geochem. Health* **2014**, *37*, 363–375. [CrossRef] [PubMed]
19. Nikonow, W.; Rammlmair, D. Risk and benefit of diffraction in Energy Dispersive X-ray fluorescence mapping. *Spectrochim. Acta Part B At. Spectrosc.* **2016**, *125*, 120–126. [CrossRef]
20. Poonoosamy, J.; Curti, G.; Grolimund, D.; van Loon, L.R.; Mäder, U. Barite precipitation following celestite dissolution in a porous medium: A SEM/BSE and μXRD/XRF study. *Geochim. Cosmochim. Acta* **2016**, *182*, 131–144. [CrossRef]
21. Rammlmair, D.; Tacke, K.; Jung, H. Application of new XRFscanning techniques to monitor crust formation in column experiments, Securing the Future. In Proceedings of the International Conference on Mining and the Environment, San Jose, CA, USA, 29 November–2 December 2001; pp. 683–692.
22. Pough, F.H.; Scovi, J.A. A field guide to rocks and minerals. *Natl. Audubon Soc.* **1996**, 2.
23. Ray, F.; Matt, W.; Wayde, M.; Loc, D. Identification of mixite minerals—An SEM and Raman spectroscopic analysis. *Mineral. Mag.* **2005**, *69*, 169–177.
24. Janaki, K.; Velraj, G. Spectroscopic studies of some fired clay Artifacts recently excavated at Tiltagudi in Talmilnadu. *Recent Res. Sci. Technol.* **2011**, *3*, 89–91.
25. Ramaswamy, V.; Vimalathithan, R.M.; Ponnusamy, V. Synthesis of well dispersed elliptical shaped barium sulphate nano particles water chloroform mixed solvent. *Arch. Phys. Res.* **2010**, *1*, 217–226.
26. Dimova, M.; Panczer, G.; Gaft, M. Spectroscopic study of barite from the Kremikovtsi Deposit (Bulgaria) with implication for its origin. *Ann. Gologiques Penins. Balk.* **2006**, 101–108. [CrossRef]
27. Aroke, U.O.; Abdulkarim, A.; Ogubunka, R.O. Fourier transform infrared characterization of kaolin, Granite, Bentonite and barite. *ATBU J. Environ. Technol.* **2013**, *6*, 1.
28. Nordstrom, D.K. What was the groundwater quality before mining in a mineralized region? Lessons from the Questa project. *Geosci. J.* **2008**, *12*, 139–149. [CrossRef]
29. Nagaraju, A. Effects of barite mine on groundwater quality in Andhra Pradesh, India. *Mine Water Environ.* **2007**, *26*, 119–123.
30. Forjanes, P.; Astilleros, J.M.; Fernández-Díaz, L. The Formation of Barite and Celestite through the Replacement of Gypsum. *Mineral* **2020**, *10*, 189. [CrossRef]
31. Bhavan, I. *Indian Minerals Yearbook-Barytes*, 54th ed.; Indira Bhavan: Kerala, India, 2015.

Review

A Systematic Review on the Application of Machine Learning in Exploiting Mineralogical Data in Mining and Mineral Industry

Mohammad Jooshaki [1,*], Alona Nad [2] and Simon Michaux [1]

1 Circular Economy Solutions, Geological Survey of Finland (GTK), 02151 Espoo, Finland; simon.michaux@gtk.fi
2 Circular Economy Solutions, Geological Survey of Finland (GTK), 83500 Outokumpu, Finland; alona.nad@gtk.fi
* Correspondence: mohammad.jooshaki@gtk.fi

Abstract: Machine learning is a subcategory of artificial intelligence, which aims to make computers capable of solving complex problems without being explicitly programmed. Availability of large datasets, development of effective algorithms, and access to the powerful computers have resulted in the unprecedented success of machine learning in recent years. This powerful tool has been employed in a plethora of science and engineering domains including mining and minerals industry. Considering the ever-increasing global demand for raw materials, complexities of the geological structure of ore deposits, and decreasing ore grade, high-quality and extensive mineralogical information is required. Comprehensive analyses of such invaluable information call for advanced and powerful techniques including machine learning. This paper presents a systematic review of the efforts that have been dedicated to the development of machine learning-based solutions for better utilizing mineralogical data in mining and mineral studies. To that end, we investigate the main reasons behind the superiority of machine learning in the relevant literature, machine learning algorithms that have been deployed, input data, concerned outputs, as well as the general trends in the subject area.

Keywords: machine learning; artificial intelligence; mineralogy; mining; mineralogical analysis

Citation: Jooshaki, M.; Nad, A.; Michaux, S. A Systematic Review on the Application of Machine Learning in Exploiting Mineralogical Data in Mining and Mineral Industry. *Minerals* **2021**, *11*, 816. https://doi.org/10.3390/min11080816

Academic Editors: Herbert Pöllmann, Uwe König and Yosoon Choi

Received: 2 June 2021
Accepted: 25 July 2021
Published: 28 July 2021

Publisher's Note: MDPI stays neutral with regard to jurisdictional claims in published maps and institutional affiliations.

1. Introduction

Artificial intelligence (AI) and machine learning (ML) have been used in a wide range of applications in the development of technology. AI is a branch of science and engineering focusing on the development of techniques to make computers capable of solving certain problems through simulating or extending human intelligence [1,2]. As a subset of AI, ML includes computational approaches aiming at extracting expertise out of experience [3,4]. In other words, the goal of ML is to use past data or known information to extract (learn) meaningful patters and associations which can be generalized to make relatively accurate predictions [3,4].

In the realm of ML, the process of leaning the past information is called *training*. Learning through training experiences to acquire new or improve previous capabilities distinguishes ML methods from traditional explicit programming of computers for performing a specific task [4–6]. Conventional programming relies on explicit modeling of a problem using the physical rules governing a specific system under study. In ML, however, the aim is to analyze the data to predict the behavior of complex systems that cannot be explicitly modeled using conventional approaches. The learning process can be either supervised or unsupervised. While in the former, the training data is labeled and the correct output is known for every instance of the past information, the latter entails the recognition of hidden patterns in the data without knowing specific outcomes a priori [5,7].

With the capability of learning from the past data or experience and generalizing that to the unseen data, ML techniques are able to solve complex problems, which cannot be effectively and efficiently addressed by the traditional methods. Such problems typically involve intricate associations among several variables influencing the system under study, fluctuating environments, and large amount of data which needs to be processed [8].

AI and ML researchers have devised a plethora of effective tools to solve the most difficult problems in computer science and engineering including speech recognition, machine vision, control of autonomous vehicles, robot control, natural language processing, medical diagnosis, climate and power demand forecasting, playing games, filtering spam emails, designing performance-based regulations based on unsupervised ML methods, and optimizing engineering problems using soft computing intelligence, to name but a handful [5,6,9–16]. Such tools have been leveraged in various industries so as to enhance the performance and efficiency. These next generation information tools, that have become more refined over the recent years, have been also applied to the mining industry—a capital-intensive business, thus, conservative and reluctant to radical changes—to improve safety, increase productivity, and reduce costs [17,18].

With the continuously increasing demand for raw materials, deeper mining, facing the complexity of the geological structure of ore deposits, and decreasing ore grade, high-quality and extensive mineralogical information is required [19–22]. The mineralogical analysis is providing critical information for calculating the duration of extraction period of a mine. Intimate knowledge of the mineralogical assembly of ores is key to understanding and solving problems encountered during exploration and mining, and during the processing of ores, concentrates, and related materials [23].

Additionally, mineral process engineering is now evolving. In the past, the practical challenges of managing and optimizing a process plant were very complex and were not effective enough to justify further development of data-based optimization. Most designs were based around empirical characterization tests, and plant operation relied heavily on operator intuition. Given the standard process operating systems, it was a challenge for the plant operators to manage the large amount of process information in a fashion where all of it was used effectively. There also is now technology available that has the capacity to revolutionize how mineral processing plants are designed and how they are operated. Instrumentation used to collect data and information has become much more sophisticated, and capable. This data can be also collected from many more positions throughout the process plant, and it can be collected at a much higher degree of resolution.

ML, a revolutionary new method of handling vast amounts of data, has been developed in the past few years to the point where it can be applied to mineral processing applications. The combination of more sophisticated instrumentation in conjunction with ML has the potential to revolutionize mineral processing standard operating practice. In the past, many mineral processing plants struggled with high variability in throughput, power draw and recovery, where most operations are designed to operate at a steady state of throughput, recovery, and metal production. Poor recovery was often observed as it happened, with limited understanding in why it was happening.

ML, if coupled with quality and timely measurement of the appropriate parameters, has the potential to diagnose the true metrics of good recovery. Time stamped measurements at multiple strategically decisive points in the plant could quantify, for example, how recovery relates to ball mill performance (under grinding or over grinding) or how cyclone performance could interact with final metal reconciliation. The true link between mineral content, mineral texture, and process performance could be quantified.

Using ML, the cause of plant variability could be isolated in real time. Depending on circumstance, this could happen soon enough to make an engineering decision, followed by operational optimization. An example of this could be using a Raman spectrometer instrument to estimate mineral content of the semi-autogenous grinding (SAG) mill feed, the results of which could be used to optimize reagent application at the flotation cells. ML could be used to focus on the best outcome. For instance, in a comminution circuit, which is the most energy and cost consuming step in mineral processing, it is proposed to use SAG Mill real-time operational variables such as feed tonnage, bearing pressure, and spindle speed in order to predict the upcoming energy consumption via ML and deep learning techniques [24]. It should be highlighted that the authors of [25] achieved impressive accuracy of 97% with the emulation of the industrial grinding circuit by the designed recurrent neural networks for the SAG mill in lead-zinc ore beneficiation process.

The potential here is that the true relationship between process units during operation could be quantified with the application of ML. The whole process flow sheet could be optimized in operation. Additionally, any given example of poor performance could be diagnosed, and the original cause could be isolated to individual process units. If this potential is realized, the next generation of mineral process practice could be developed.

ML algorithms such as artificial neural network (ANN), support vector machine (SVM), regression tree (RT), and random forest (RF) are powerful data driven methods that are becoming extremely popular in such applications as the mapping of mineral prospectivity [26–28], mapping geochemical anomalies [29–31], geological mapping [32–35], drill-core mapping [36–38], and mineral phase segmentation for X-ray microcomputed tomography data [39–41].

Inspired by these remarks, this paper aims to systematically survey the relevant literature for the sake of investigating what has been carried out in the realm of enhanced exploitation of mineralogical analysis data in mining and minerals industry. In a systematic review, the body of knowledge on a specific subject is investigated to answer a set of predetermined questions in such a way that the data and methods used are definite. Given the sufficient details provided in a systematic literature review, the users can more conveniently determine its trustworthiness and the usefulness of the statistics, discussions, and findings it provides [42].

To be more specific, areas examined in this paper include mainly the problems in the field of utilizing mineralogical analysis data for mining and minerals industry that have been solved using ML techniques. The reason behind using ML in such studies, ML tools that have been developed and applied, data inputs, required outputs, as well as the main trends in this subject area are also assessed in this paper.

The remainder of this paper is organized as follows. In Section 2, the main research questions that we aim to answer in this review as well as the search method, information source, and the selection criteria are explained. Section 3 provides answers to the research questions, and lastly, Section 4 concludes the paper.

2. Main Research Questions and Review Methods

As pointed out before, in a systematic review, the main objective is to investigate the body of knowledge to address a set of research questions [42,43]. This must be done using concrete methods and procedures for the sake of transparency and ease of evaluating the objectivity and trustworthiness of the figures and outcomes reported for the potential readers [42].

Considering the groundbreaking advances and flourishing developments in the area of ML, it has been leveraged in various fields to unprecedentedly solve complex problems that could not be tackled via conventional methods effectively. Given the crucial role of mineralogical monitoring at every stage of minerals industry value-chain, from geoscience research and exploration phases to the final processing, and the complexities associated with exploiting valuable information out of mineralogical data, we aim to investigate the steps taken towards adopting ML in this area. In other words, this review focuses on the applications of ML for enhancing and facilitating the mineralogical monitoring and the utilization of mineralogical data in mining and minerals industry. To that end, the main questions that this review aims to answer comprise:

(1) What problems in the area of mineralogical studies for mining industry have been addressed using ML techniques in the existing literature?
(2) Why the use of ML in such applications is required?

(3) What are the outputs predicted or modeled using ML in those problems? What input parameters have been used?

(4) What ML methods have been employed?

(5) What are the general trends in the area under study?

The purpose of the first two questions is to investigate the sort of problems in the area under study for which the use of ML techniques is advantageous and their specific complexities that favor ML over traditional approaches. Considering that ML techniques are typically used to find useful associations in the data to predict specific outputs in the case of supervised learning or to determine significant patterns among input features in an unsupervised setting, the third question intends to explore the input and output variables employed in the existing literature. This can be particularly useful for understanding potential data that needs to be collected for estimating a specific set of desired output variables. Complementary to the previous question, the fourth question focuses on the methods used to model the relationships between input and output variables. Lastly, the final question will address the general trends in the application of ML in mineralogical monitoring in mining and minerals industry.

2.1. Information Source and Search Strategy

We utilized Scopus [44] as the search engine for finding the relevant publications. Based on our review objectives, we considered three main tiers of keywords, namely target modeling approach, analysis, and industry, each including few relevant keywords as depicted in Figure 1. As illustrated, using the logical operators available in Scopus search tool, we set a query so as to search through the records and reach the publications that include at least one of the keywords in each tier in their title, abstract, or list of keywords. In order to have an estimate of the early works published in the subject area, we did not filter any record based on the publication date in our search query. Nonetheless, as we will discuss in the next section, the works published before 2000 were disregarded during the selection procedure. It is worth mentioning that we set no limitations on the search source, thereby the search query was applied to all the records covered by Scopus—the largest database of peer-reviewed literature [44].

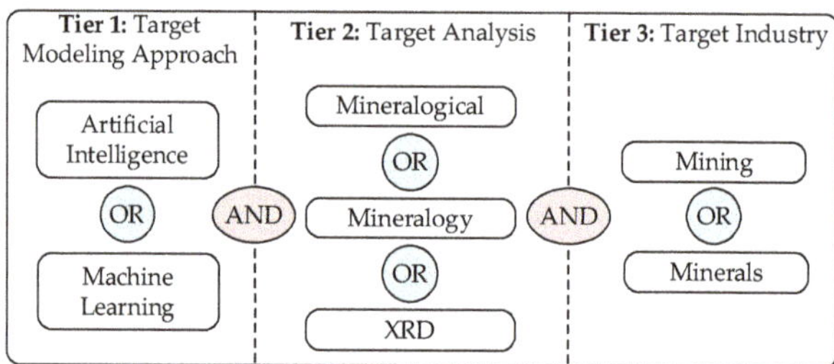

Figure 1. Search keywords and search query logic.

2.2. Eligibility Criteria

Based upon the search method explained in the previous section, we reached 145 candidate scientific publications. As depicted in Figure 2, in the first step, we discarded 9 manuscripts including non-English papers and those published before the year 2000. In the next stage, the review articles and conference reviews were disregarded, a total of 8 records. The remaining 128 records were all sequentially evaluated by each of the authors to select the publications relevant to this review. The evaluation entailed reading the title, abstract, and list of keywords of the records and skimming through the papers if required. Lastly, 55 papers were selected to be included in the review and answer the questions. It is worth mentioning that the publications excluded in the last stage comprise different topics not directly relevant to the mining and minerals industry.

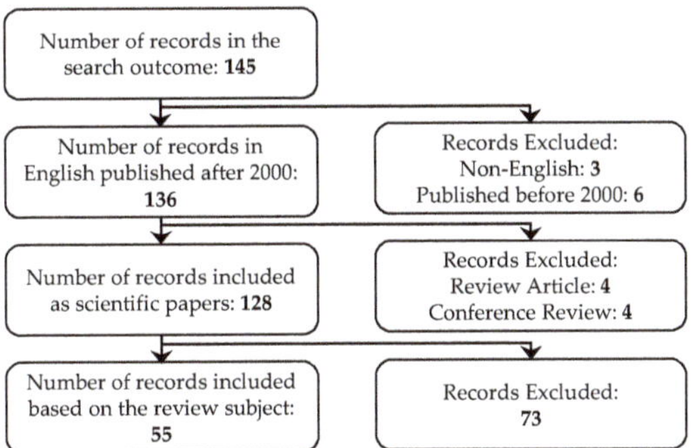

Figure 2. Process of screening the search results.

Figure 3 shows the share of the topics excluded at the last stage of the selection procedure. As per this figure, the majority of the papers excluded are related to petroleum and gas industry, soil, and space exploration. About a quarter of the records excluded is on environmental studies, works in which ML techniques are not employed, and miscellaneous topics such as metallurgy, recycling in cement industry, archaeometry, geophysics, medicine, and sedimentology.

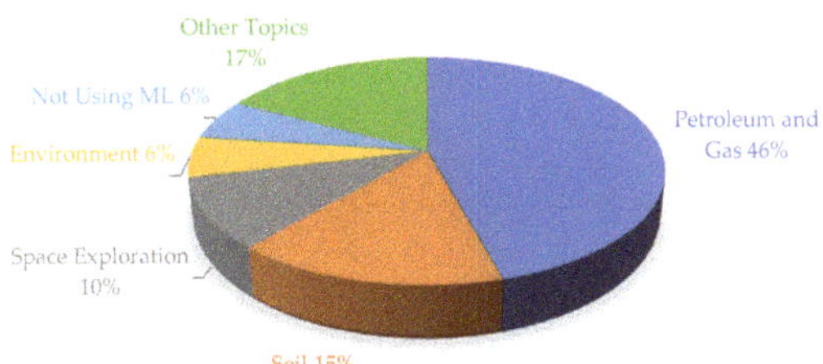

Figure 3. Share of the topics excluded from the review in the last stage of the selection process.

Title of the journals with more than one record in the list of selected records are provided in Table 1.

Table 1. Title of journals with more than one entry in the final list.

Journal Title	Number of Selected Records
Minerals Engineering	6
Journal of Geochemical Exploration	6
Computers and Geosciences	4
Applied Geochemistry	4
Ore Geology Reviews	3
Lecture Notes in Computer Science	3
Remote Sensing	2

3. Results

In this part, the main research questions stated in Section 2 are addressed based upon the assessment of the selected works.

3.1. Problems in the Selected Studies Addressed Using ML Techniques

During the last decade, the number of available multi-parameter datasets in the mining industry increased rapidly as a result of applying advanced technologies to assist in the exploration process. To integrate and handle such large datasets, special tools are required. One such tool is the ML, which is well suited and proved to be promising for tackling the problem of mapping geochemical anomalies [45–47] and mineral prospectivity, due to its ability to effectively integrate and analyze large geoscience datasets [48–53]. ML and AI are actively used for mining complex, high-level, and nonlinear geospatial data and for extracting previously unknown patterns related to geological processes [45]. These techniques were applied in the identification of mineralization related geochemical anomalies in China [45,47,54–56], as well as generating a prospectivity map for targeting gold mineralization in Canada [49,50] and China [51], for the detection of iron caps in Morocco [57], for creating a continuous mineral systems model for chromite deposits in Iran [58], and geological mapping studies using the characteristics of rocks [59,60]. The authors in [61,62] integrated multi-sensor remote sensing techniques such as drone-borne photography and hyperspectral imaging for processing with ML algorithms in order to generate the geological mapping.

Another important field, where the analysis of the hyperspectral imaging has been applied is drill-core mapping. It is well known that drilling is a decisive step for validating and modeling ore deposits. The hyperspectral imaging technique provides a rapid and non-invasive analytical method for the core samples in terms of mineralogical characterization [63]. Recently, ML techniques have been suggested for automating the process of mineral mapping based on drill-core hyperspectral data [63–68]. However, several obstacles might occur due to the small amount of representative data for training purposes. To tackle this problem, resampling and co-registration procedures for the high-resolution mineralogical data obtained by the scanning electron microscope (SEM)-based mineral liberation analysis (MLA) of the hyperspectral data was implemented in [63,64,67]. The new co-registered data was used for training purposes through a classification algorithm. Mainly, the RF classifier is used due to its high performance when small training samples are available [64,67]. Nevertheless, in [63], three methods, namely RF, SVM, and ANN were employed for the classification and regression tasks. The authors reported that the RF is more robust to unbalanced and sparse training sets.

Mineral processing should always be considered in the context of geological, mineral assemblage, and texture of ores in order to predict grinding and concentration requirements, feasible concentrate grades, and potential difficulties of separation [69]. A promising technique was proposed by the authors of [70] in the context of control of mineral processing plants for the identification of minerals in slurry samples through multispectral image processing. The study was focused on the base metal sulfides minerals and the main goal was to develop set-up aims to enable the measurement of specular-like reflections on the surface of the particles. A supervised classification approach has been used to process the acquired data. In [71], the mineralogical composition of the final products (copper concentrates) was analyzed by a near-infrared hyperspectral camera. ML has been used to provide the mineralogical spatial distribution of the different components in the samples through the analysis of the reflective images.

The application of X-ray microcomputed tomography (μCT) in the mineral industry has been growing due to its noninvasive nature of sample analysis. X-ray microtomography allows achieving high-resolution images with pixel sizes in the micrometer range. However, the grayscale values of mineral phases in a sample should be different enough to be segmented. Despite the fact that the manual segmentation of those images made by a highly experienced specialist is one of the best methods for segmentation, the process is highly time-consuming. Moreover, the procedure of preparing polished thin-section for microscope is long and the number of core samples is limited. As a consequence, the main challenge in using ML is the limited number of ground-truth (or segmented) images that are available for the training step. For instance, in [72], only 20 images were manually segmented to be analyzed by a convolutional neural network (CNN), thereby resorting the authors to employ data augmentation techniques. The authors in [73] have applied supervised and unsupervised methods to the training data obtained by the matching method for back-scattered electron (BSE) mineral map to its corresponding μCT slice for one drill core sample. Classifying voxels in X-ray microtomographic scans of mineralogical samples is another problem that has been solved by applying the ML techniques [74].

The observation of optical properties of a mineral in a polarized microscope rotation stage is a commonly used method for mineral type classification. This task can be automated by the application of digital image processing techniques and AI technologies [75–78].

Interesting solutions by the implementation of an ML methodology to the prediction of material properties from the nepheline syenite deposit was discussed in [79]. The challenges with calculating the amphibole formula from electron microprobe analysis can be solved by applying ML [76]. Another problem in mineralogy study that can be addressed directly via deep learning algorithms is differentiation of quartz from resin in optical microscopy images of iron ores [80].

3.2. The Main Resaons behind Using ML in the Selected Studies

ML techniques are typically employed to solve problems for which the application of traditional approaches is either impossible or very sophisticated [8]. Such problems might entail typical tasks that human beings or animals can perform routinely, yet the process of doing such tasks is relatively unknown, tasks that involve processing an excessive amount of data with complex unidentified relationships and patterns, or tasks that require interaction with constantly changing environments such that high levels of adaptivity are required [4]. Our review of the selected papers revealed that, albeit all these three reasons can account for the necessity of using ML techniques in the area under study, the second category of tasks is more common. In other words, in most of the studies investigated, the researchers attempted to leverage ML techniques to cope with large and complex datasets.

Let us take mineral exploration as an example, new mineral prospect or deposit targets are deeper, thus, more difficult to find [81]. Therefore, it becomes of utmost importance to predict, relatively accurately, regions with higher potentials for new deposits based on the large datasets of various types of measurements. The dataset can contain lithogeochemical [49,82], spatial [49,50], geochemical [45,55,81,83], geophysical [81], geological [81], concentration of indicator elements [47,51,52,54,56,65,68], hyperspectral [57,60,61], spatial proxies [58], total magnetic intensity [52], isostatic residual gravity [52] data. It is worth emphasizing that in most of such studies mineralogical analyses results are either used to generate the input features for the ML models or ground truth for training such models. Obviously, analyzing such massive and complex datasets is challenging, adding to that the nonlinearities and hidden interdependencies and patterns among different features. This calls for deploying multivariate ML models to effectively explore the data and attain valuable insights.

In some applications, especially the tasks entailing image processing, ML is proposed to automate manual operations to enhance productivity via enhancing speed and reducing human errors. As an example, the authors in [78] proposed a technique for the identification and classification of hematite crystals in iron ore using optical microscopic images. Presence of high noise can also result in the ineffectiveness of the conventional techniques, thereby giving rise to the application of ML. For instance, extracting quantitative mineralogical information about composition, porosity, and particle size through processing X-ray microtomography scans of ore samples can be quite challenging due to the presence of noise [74].

Another important driving force for the deployment of ML is the cost reduction. In [47], ML is leveraged to select a small set of indicator elements to detect chemical anomalies with the main goal of avoiding the unnecessary cost of element concentration measurements for mineral exploration. To save time and money through reducing the number of samples on which X-ray diffraction (XRD) measurements must be obtained, the authors in [84] proposed an artificial neural network-based model for estimating the mineralogical compositions based upon the elemental data from X-ray fluorescence (XRF) instruments. In a conceptually similar manner, ML is used in [85] to reconstruct synthetic 3D models of porous rocks from 2D images of thin sections.

3.3. The Outputs Predicted/Modeled Using ML in the Selected Literature and the Inputs Utilized

Based on the reviewed literature it can be considered that the most widely used input data for analyzing by ML in the mining and mineral industry is a set of *digital images*. As some examples, the hyperspectral data was used for discrimination of lithologic domains in geology mapping [61], and a combination of the multispectral, RGB, and hyperspectral data was analyzed by ML algorithms to create a digital outcrop model for precise geology mapping [62]. Moreover, the hyperspectral imagery has been used for classifying rock type and mineralogy [86], for predicting the presence of specific minerals [64] or mineral abundance [63] in drill-core samples, as well as for drill-core mineral mapping [67,68] and mapping of mine face geology [53]. In [87], a three-stage method is proposed for the segmentation of hyperspectral images with the main goal of preparing the data required for the classification of such images. In order to curb the noise in the spectral domain, Gaussian processes (GP)—a type of supervised ML model—are used in [88] as a preprocessing step before extracting the mineralogical information from the images.

The authors in [89] have discriminated a rock texture information through image processing and machine learning algorithms by studying a geologist-labeled digital photograph database from drill-hole samples. The main contribution of this work is "a novel texture characterization technique to compare image textures of drill-hole samples and discriminate between different rock texture classes".

The study [41] used the association indicator matrix (AIM) and local binary pattern (LBP) texture analysis methods to get quantitative textural descriptors of drill core samples with relatively high accuracy of 84% and 88%, respectively, for AIM and 3D LBP. An automatic method for the classification of hematite textures in Brazilian iron ores according to their textural types through applying an AI technique for analyzing the images from a reflected light microscope and a digital camera is described in [78]. New optical properties have been extracted from the digital images acquired under cross and plane-polarized light from different rock thin sections. ML was deployed for mineral classification by analyzing the optical properties of color and texture of a pair of images of the same mineral taken on different lights [77].

Deep learning and ML have produced accurate results in different applications when various images are available for the training [72]. The researchers in [72,73] proposed implementing ML algorithms to enhance automatic segmentation of mineral phases based on the analysis of the images from the X-ray microcomputed tomography (μCT). However, it should be noticed that acquiring μCT images is expensive and time-consuming, which affects the limited available dataset. Therefore, a supervised ML algorithm in which the user pre-defines the underlying pattern of the data, and the computer system builds a prediction model based on the pre-defined pattern (training data) [73] could be successfully applied to tackle this problem even with a small number of images. The supervised classification method was used for generating a 2D mineral map of chromite sample from optical microscopic images [90].

Alongside the image analysis, other input features among the mineralogical study for the mining and minerals industry have been addressed using ML techniques. The dataset containing geochemical data was used for extracting features related to mineralization via a deep learning algorithm and these features were then integrated as an anomaly map [45,47,54,56]. Applying the deep learning algorithms as a subcategory of machine learning algorithms can lead to improving the accuracy of classification or prediction by replacing the manual selection [45]. Such techniques have been employed in recognizing geochemical anomalies related to mineralization via deep autoencoder networks [46], deep variational autoencoder network [45], convolutional autoencoder networks [91], and combining deep learning with other anomaly detection methods [54,56].

The bulk chemistry data from the mining company open-pit database was used as an input for the prediction of laboratory concentrate yield and modal mineralogy for the nepheline syenite deposit in Norway by adopting a neural network approach [79]. The data collected by the electron probe microanalyzer (EPMA) was analyzed with an ML method aimed to be established for calculating the amphibole formula [76].

The lithogeochemical data of sandstones from diamond drill cores [82] and lithogeochemical major oxide data from the Swayze greenstone belt [49] have been used for the identification of sandstones above blind uranium deposits through an ML technique in a first case and for modeling of orogenic gold prospectivity mapping by deploying a support vector machine and an artificial neural network in the second one.

3.4. The ML Methods Leveraged in the Selected Works

ML methods are typically categorized based on different aspects. From the standpoint of the learning type, they are generally regarded as supervised and unsupervised approaches [3,4]. Other classes, namely semi-supervised and reinforcement learning, are also available [3], yet we found no instances for the applications of these methods in the reviewed papers.

In supervised learning, the data is labeled with the correct outputs such that during the training process, the model can understand the underlying associations between input features and output variables. Moreover, for testing the performance of the algorithm, predictions of the trained model for test examples can be benchmarked against the known labels to estimate the accuracy of the resulting model. In stark contrast, unsupervised learning entails unlabeled data from which the learner must find meaningful patterns [3]. In this setting, it can be challenging to estimate the performance of the model [3].

As shown in Figure 4a, the majority of the methods used in the reviewed papers fall in the supervised learning category. More precisely, among the 17% of the reviewed works that used unsupervised learning models, only 6% solely leveraged unsupervised learning [48,55,83], but in the remaining 11%, a combination of the supervised and unsupervised learning techniques is utilized [54,61,73,79,92,93]. In such works, unsupervised learning methods are typically used for the feature extraction and preprocessing of the data to be used in a supervised learning process.

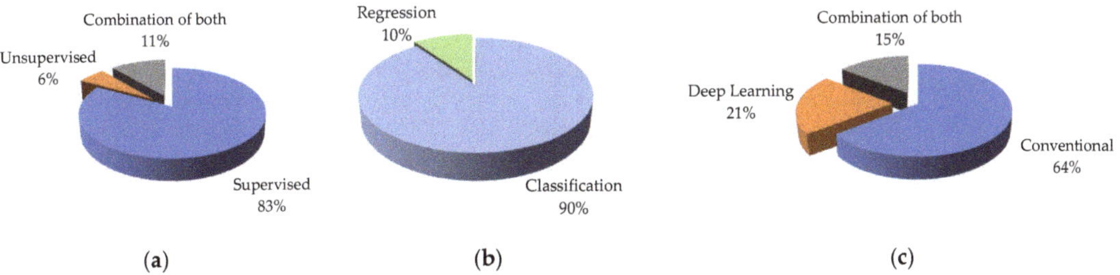

Figure 4. Type of ML techniques employed in the reviewed papers: (**a**) type of learning; (**b**) type of problem; (**c**) type of ML model.

The task of supervised learning can be either classification or regression [8]. In a classification task, the labels are categorical, i.e., have a set of limited values, however, in the case of regression, labels take continuous numerical values. As per Figure 4b, in 90% of the reviewed records, the objective of using ML is classification. Classification tasks can be, for instance, determining the type of minerals [73,77,94,95], texture [66,89], or rock [86,93], class of mineral face [86], class of hematite crystals [78], distinguishing between quartz and resin in optical microscopy images [80], presence or absence of specific minerals in a sample [96], zeolite type [97], material fingerprints [98], class of regolith landform [99], and class of carbonates [100] based on a set of measurements or known features about a material. On the other hand, in regression problems, the goal is to estimate a continuous numerical value, for instance, prediction of concentrate yield and modal mineralogy [79] and estimation of drill-core mineral abundance [63], mineral density of elements [101], and calculation of amphibole formula [76].

From another perspective, ML methods can be categorized as conventional models and deep learning techniques. The latter is based upon the multilayer artificial neural networks. As presented in Figure 4c, deep learning models have been used in 36% of the reviewed papers from which 21% solely utilized deep learning, whereas the remainder employed both deep learning and conventional machine learning methods to choose the best [49,54,58,63,66,70,96]. It is worth noting that the superiority of an approach depends on the application.

A list of different ML methods used in the reviewed papers together with their frequency of usage is provided in Table 2. As per this table, SVM, RF, and different types of feed-forward ANN are the most frequently used approaches in the supervised learning category. SVM classifiers are very powerful and flexible for linear and nonlinear classification of complex but relatively small-sized datasets [8]. The main idea behind a linear SVM classification is to find an optimal hyper plane that can separate different classes while maximizing the margin of the plane [102]. In the case of nonlinearly separable datasets, the data is mapped to a higher dimension space, where the classes become linearly separable [102]. The mapping is carried out using a kernel function, typically a polynomial or Gaussian radial basis function (RBF) [8]. In most of the reviewed works, SVM is used with a Gaussian RBF kernel.

RF belongs to a category of ML named *ensemble methods*. Ensemble methods are based on the *wisdom of crowd* concept, implying that aggregating the outputs of numerous simple models through a voting system usually performs better than leveraging a single but more complex model [8]. An RF comprises several classification and regression trees, each of which are trained on a bootstrap sample of the original dataset [103]. Notwithstanding their simplicity, RFs are among the most powerful ML techniques [8,104].

ANNs have a relatively long history and were originally developed to simulate the nervous system. An ANN is comprised of numerous basic units called artificial neurons. From a mathematical perspective, an ANN is a complex nonlinear function, which can be tuned for a specific task to perform the desired mapping from an input vector to the output value(s) [104]. ANNs proved to be very powerful tools and outperformed the other ML algorithms in many applications [104].

On the other hand, in the class of unsupervised learning techniques, K-means and hierarchical clustering are used more frequently compared to the other techniques. K-means algorithm partitions datapoints into a predetermined number of clusters such that the similarity among the points in a cluster is the highest, while it is the lowest for the datapoints falling in different clusters. To achieve this goal, in K-means method, an optimization model is solved to minimize the sum of the distances of the datapoints to the nearest cluster center, where the positions of the cluster centers are the decision variables of the optimization model [105]. In contrast to the K-means algorithm, which is centroid-based, i.e., assigning a datapoint to the cluster with the nearest cluster center, in hierarchical clustering datapoints with distances lower than a specific threshold are assigned to the same cluster [106].

Aside from the ML techniques presented in Table 2, the implementation of real-time expert systems in mineral processing operations is discussed in [107], where generating quantitative data using natural language processing (NLP) of process data including ore mineralogy is proposed.

Table 3 summarizes the applications of ML methods as well as the type of datasets utilized in the reviewed papers. As per this table, principal component analysis (PCA) is the most frequently used technique for feature engineering, more specifically for dimensionality reduction [48,49,56,67,76,82,83,95,96]. Weight of evidence (WOE) [49], minimum noise fraction (MNF) [61], orthogonal total variation component analysis (OTVCA) [61,62,65], stacked denoising autoencoder (SDAE) [56], hierarchical clustering [56], CNN [72,80], grey-level co-occurrence matrix (GLCM) statistics [66], local binary patterns (LBP) [66], maximum margin metric learning (MMML) [55], and K-means++ [93] are the other techniques leveraged in the selected works for feature engineering.

Table 2. ML models leveraged in the selected papers.

Category	ML Method	Reference
Supervised	Support vector machine (SVM)	[49,52,54,57,59,61–63,65,66,68,86,87,92]
	Random forest (RF)	[47,58,63,64,66–68,73,82,87,90–93,97]
	Feed-forward artificial neural network (FF-ANN)	[58,63,66,70,79,84,87,94–96,99]
	k-nearest neighbors (k-NN)	[66,71,73,77,87,89]
	Convolutional neural network (CNN)	[50,51,72,80,85]
	Gaussian processes (GP) [1]	[53,60,86,88]
	Decision tree (DT)	[75,77,87,96]
	Linear discriminant analysis (LDA)	[70,78,82,92]
	Radial basis function neural networks (RBFNN)	[49,81]
	Adaptive Coherence Estimator (ACE)	[55]
	Bayes nets	[100]
	Isolation forest (IF)	[56]
	Linear regression (LR)	[101]
	Naive Bayes (NB)	[78,87]
	Principal components regression (PCR)	[76]
	Quadratic discriminant analysis (QDA)	[92]
	Support vector regressor	[101]
	Variational autoencoder (VAE) network	[45]
Unsupervised	K-means clustering	[48,73,92,93]
	Hierarchical clustering	[48,79,92]
	Deep belief networks(DBNs)	[54]
	Fuzzy C-means clustering	[73]
	Gaussian mixture model (GMM) clustering	[48]
	Unsupervised random forest	[83]

[1] With either the squared exponential (SE) or the observation angle dependent (OAD) covariance functions.

3.5. General Trends and Research Gaps in the Application of ML in the Selected Literature

Development of effective methods together with the availability of large datasets and more powerful hardware have resulted in flourishing of ML in recent years [6,8]. This has been reflected in the application of ML in various science and engineering domains. Figure 5 shows the yearly distribution of reviewed literature and their type, namely research articles and conference papers. As per this figure, the number of publications has increased rapidly since 2018. It is worth mentioning that 9 out of 55 papers reviewed are open access.

Table 3. ML models leveraged—application and dataset.

Application	Dataset	Feature Engineering Method	ML Technique
Calculating amphibole formula	Routine electron microprobe analysis (EMPA) data	PCA [76]	PCR [76]
Characterizing the composition of igneous rocks	Raman spectra of mineral samples	PCA [96]	DT [97]; ANN [96]
Classification and prediction of alteration facies	Multi-element geochemical data	Hierarchical clustering [92]; K-means [92]	SVM [92]; LDA [92]; QDA [92]; CART [92]; RF [92]
Classification of inorganic solid materials of known structure	Topological attributes of Delaunay simplex properties	_1	RF [97]
Classifying hematite crystals	Optical microscope images	LDA [78]	NB [78]
Detecting potential Cu mineralization in bedrocks based on the composition of basal till	Geochemical data	PCA [83]	Unsupervised RF [83]
Determining mineralogical spatial distribution of the different components in a concentrate sample	Near-infrared hyperspectral image	-	k-NN [71]
Determining type of rock texture	Rock images	-	k-NN [89]
Discrimination of lithologic domains	Hyperspectral data	MNF [61]; OTVCA [61]	SVM [61]
Drill-core mapping	Hyperspectral data	PCA [67]	RF [64,67,68]; SVM [68]
	Hyperspectral short-wave infrared (SWIR); scanning electron microscopy-based image analyses using a mineral liberation analyzer (SEM-MLA); visible/near-infrared (VNIR); long-wave infrared (LWIR)	OTVCA [65]	RF [63]; SVM [63]; [65]; FF-ANN [63]
Estimating the mineralogical compositions	Elemental data acquired using X-ray fluorescence (XRF) instruments	-	ANN [84]
Finding association between imaging and XRF sensing	Images of rock samples	-	LR [101]; SVM [101]
Generating 2D mineral map of chromite samples	Optical micrograph images	-	RF [90]
Geochemical anomaly detection; prospectivity for future exploration	Geochemical exploration data; concentration of major and trace elements	Feature elimination with cross-validation based on random forest [47]; unsupervised deep belief networks (DBNs) [54]; MMML [55]; hierarchical clustering [56]; SDAE [56]; PCA [56]	VAE [45]; RF [47]; CNN [51]; SVM [54]; ACE [55]; IF [56]
	Litho geochemistry of sandstones obtained from drill cores	PCA [82]	LDA [82]; RF [82]
	Geochemical assay (ppm Cu); total magnetic intensity; isostaticresidual gravity	-	SVM [52]
	Spatial proxies	-	ANN [58]; RF [58]

Table 3. Cont.

Application	Dataset	Feature Engineering Method	ML Technique
Geochemical imaging	Qualitative LIBS spectral data	PCA [48]	K-means [48]; agglomerative hierarchical clustering [48]; GMM [48]
Geological and Geophysical Mapping for mineral exploration, mine planning, and ore extraction	Multispectral, RGB, and hyperspectral data	OTVCA [62]	SVM [62]
Geological texture classification	Images of drill cores	GLCM [66]; LBP [66]	RF [66]; SVM [66]; k-NN [66]; ANN [66]
Identifying and mapping geology andmineralogy on a vertical mine face	Hyperspectral data	-	GP [60]
Mapping of gold deposits and prospects	Lithogeochemical major oxide data; spatial data	PCA [49]; WOE [49]	RBFNN [49]; SVM [49]
	Geoscience data	-	CNN [50]
	Geological, geochemical, structural, and geophysical datasets	-	RBFLN [81]
Mineral identification	Reflected light optical microscopy (RLOM) images	-	CNN [80]
	μCT images	-	CNN [72]
	Multispectral images	-	LDA [70]; FF-ANN [70]
	Reflectance spectra	-	Bayes nets [100]
	X-ray spectrum data	PCA [95]	ANN [95]
	X-ray microtomography scans	-	Fuzzy inference system (FIS) [74]
	Images of microscopic rock thin section (RGB pixels)	-	k-NN [77]; DT [77]
Optical identification of minerals	Mineral properties such as color, hardness, pleochroism, anisotropism, and internal reflections	Cramer's Vand Pearson correlation coefficient (PCC) [75]	DT [75]
Prediction of concentrate yield and modal mineralogy	Bulk chemistry data from the mining company open pit database	-	ANN [79]
Predicting rock type and mine face, detecting hydrothermal alteration	Physical properties of rocks	-	SVM [59]
	Hyperspectral data	-	GP [53,86]; SVM [86]; SAM [86]
	Multi-element geochemistry	K-means++ [93]	RF [93]
	Images of the rocks	-	ANN [94]
Reducing noise in hyperspectral data	Hyperspectral imagery from vertical mine faces	-	GP [88]
Regolith landform mapping	Gamma-ray spectrometry data; derivatives of the SRTM elevation model, Landsat, and polarimetric radar	-	ANN [99]

Table 3. *Cont.*

Application	Dataset	Feature Engineering Method	ML Technique
Segmenting hyperspectral images	Hyperspectral data	-	SVM [87]; RF [87]; ANN [87]; k-NN [87]; DT [87]; NB [87]
Segmenting mineral phases	μCT dataset	-	K-means [73]; FCM [73]; RF [73]; k-nearest neighbors [73]

[1] Feature engineering methods are not used/mentioned in the corresponding works.

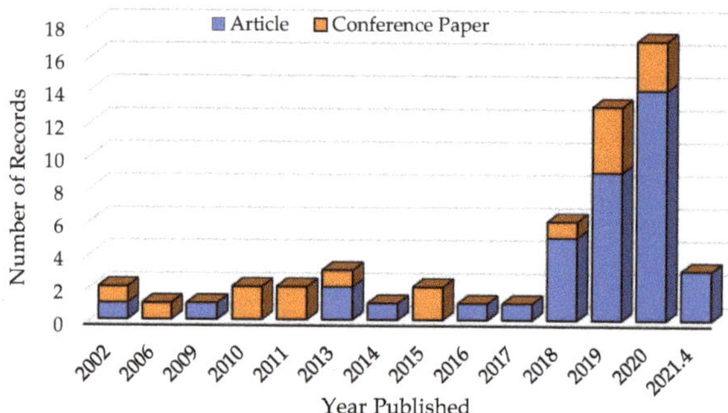

Figure 5. Yearly distribution of the number of selected papers (January 2000–April 2021).

For the sake of providing a general insight into the scope of the reviewed papers, Figure 6 illustrates the word frequency map of their titles, where the more popular words are represented with a bigger font size. As depicted, aside from the common keywords, namely *using*, *machine*, *learning*, and *mineral*, we can generally infer the main trends discussed in the previous sections considering the relatively high frequency of the words *hyperspectral*, *mapping*, and *classification*. Complementary to this, Figure 7 presents the word frequency map for the authors' keywords as well as index keywords, where aside from the familiar words, the terms *exploration*, *geochemical*, and *mapping* are perceptible. These, in line with the findings presented previously, represent the current trends in the application of ML for processing large datasets comprising mainly hyperspectral and geochemical data for anomaly detection and mapping of minerals in exploration studies.

Based upon our review of the selected works, we noticed the lack of high-quality data for applying ML in the mining and mineral studies. This issue, in many cases, is not simply related to not storing the data, but the unavailability of accurate and reliable labels for the data, which is required for training the supervised learning models. In many cases, the required labels need to be generated manually, thereby it is not only time consuming but also prone to biases and human errors. Unfortunately, since most of the works lack economic studies about the practical value of leveraging ML in the proposed applications, investing in providing reliable datasets seems to be challenging. Thus, a potentially valuable research avenue is the economic evaluation of using AI and ML in mining and mineral industry.

Figure 6. Word frequency map of the title of the reviewed records.

Figure 7. Frequently used words in the author and index keywords of the reviewed papers.

Considering the case dependency of the ML models and lack of sufficient training data, potential benefits of the *transfer learning* concept can be evaluated in the future works to curb such issues. It is worth emphasizing that in the reviewed works, we found examples of using transfer learning, yet they are limited to utilizing the models trained for general applications such as conventional image classifiers in special domains, e.g., geochemical anomaly identification [51]. However, the possibility of benefiting from a model trained on a specific dataset in a different case, e.g., using a model trained for mapping gold deposits in an area to facilitate the development of a new model for another geographic location, needs further evaluation.

Another point that is typically overlooked in the studied literature is the importance of feature engineering, especially in the case of conventional ML techniques, and hyperparameter setting. As the success of ML methods is highly sensitive to the selection of the input features as well as hyperparameters of the models [10], it is always beneficial to assess different sets of input features as well as hyperparameters to reach a more accurate model.

In addition, to the best of our knowledge, the application of semi-supervised techniques and reinforcement learning is missing in the existing literature on the application of ML in exploiting mineralogical data in mining and mineral industry. Such techniques can prove valuable especially in the reviewed applications where the judgement of specialists is always required considering the criticality of the tasks and that the frequency of unexpected cases where the ML models cannot generalize well may be relatively high. In this regard, a potential solution could be the application of a semi-supervised learning method on top of the current supervised techniques to decide whether the outcomes are reliable or further analysis and evaluations need to be carried out for specific samples.

4. Conclusions

In this paper, a systematic review of the works carried out on the application of machine learning for innovative use of mineralogical information in mining and mineral studies was presented. The search strategy resulted in a total of 145 records from which

55 publications were carefully chosen following the presented selection criteria. The selected papers were then thoroughly investigated to answer the main research questions involving (1) the types of problems in the area under study for which applying ML techniques are advantageous, (2) the specific complexities that favor ML over traditional approaches in such problems, (3) the most common type of datasets and output variables employed in the existing literature, (4) the ML algorithms leveraged, and (5) general trends in the area under investigation.

The review results revealed that the ML techniques have been used in a wide range of applications including geochemical anomaly mapping, mineral prospectivity, drill-core mapping, mineral processing, segmentation of μCT images, prediction of material properties, and calculating the amphibole formula, to name but a few. Analyzing massive and complex datasets with nonlinearities and hidden underlying interdependencies and patterns among different features, automating manual operations to improve productivity via enhancing speed and reducing human errors, cost reduction, and dealing with the problems caused by high noise are among the most significant reasons behind using ML in such studies. The main datasets used in these studies comprise hyperspectral images, μCT images, optical microscopic images, geochemical data, lithogeochemical data, data collected by the electron probe microanalyzer, as well as spatial, geophysical, geological, total magnetic intensity, and isostatic residual gravity data. In addition, support vector machine, random forest, and artificial neural networks are concluded to be the most frequently used supervised learning algorithms, whereas K-means clustering and hierarchical clustering are among the unsupervised learning models used more in the selected literature.

Author Contributions: Conceptualization, S.M. and M.J.; methodology, M.J.; software, M.J.; validation, M.J., S.M., and A.N.; formal analysis, M.J.; investigation, M.J., A.N., and S.M; resources, M.J.; data curation, M.J.; writing—original draft preparation, M.J. and A.N.; writing—review and editing, M.J. and S.M.; visualization, M.J.; supervision, M.J. and S.M.; project administration, S.M.; funding acquisition, S.M. All authors have read and agreed to the published version of the manuscript.

Funding: This research was funded by Geological Survey of Finland (GTK).

Data Availability Statement: Not applicable.

Acknowledgments: In this section, you can acknowledge any support given which is not covered by the author contribution or funding sections. This may include administrative and technical support, or donations in kind (e.g., materials used for experiments).

Conflicts of Interest: The authors declare no conflict of interest.

References

1. Ertel, W. *Introduction to Artificial Intelligence*, 2nd ed.; Springer: Berlin, Germany, 2018; pp. 1–20.
2. Deng, L. Artificial intelligence in the rising wave of deep learning: The historical path and future outlook [Perspectives]. *IEEE Signal Process. Mag.* **2018**, *35*, 177–180. [CrossRef]
3. Mohri, M.; Rostamizadeh, A.; Talwalkar, A. *Foundations of Machine Learning*, 2nd ed.; MIT Press: Massachusetts, MA, USA, 2018; pp. 1–7.
4. Shalev-Shwartz, S.; Ben-David, S. *Understanding Machine Learning: From Theory to Algorithms*; Cambridge University Press: Cambridge, UK, 2014; pp. 1–8.
5. Alzubi, J.; Nayyar, A.; Kumar, A. Machine learning from theory to algorithms: An overview. *J. Phys. Conf. Ser.* **2018**, *1142*, 012012. [CrossRef]
6. Jordan, M.I.; Mitchell, T.M. Machine learning: Trends, perspectives, and prospects. *Science* **2015**, *349*, 255–260. [CrossRef] [PubMed]
7. Alpaydin, E. *Introduction to Machine Learning*; MIT Press: Massachusetts, MA, USA, 2020; pp. 1–13.
8. Géron, A. *Hands-on Machine Learoing with Scikit-Learn, Keras, and TensorFlow: Concepts, Tools, and Techniques to Build Intelligent Systems*; O'Reilly Media: California, CA, USA, 2019; pp. 1–33.
9. Richens, J.G.; Lee, C.M.; Johri, S. Improving the accuracy of medical diagnosis with causal machine learning. *Nat. Commun.* **2020**, *11*, 3923. [CrossRef]
10. Taheri, S.; Jooshaki, M.; Moeini-Aghtaie, M. Long-term planning of integrated local energy systems using deep learning algorithms. *Int. J. Electr. Power Energy Syst.* **2021**, *129*, 106855. [CrossRef]

11. Jooshaki, M.; Abbaspour, A.; Fotuhi-Firuzabad, M.; Moeini-Aghtaie, M.; Lehtonen, M. Designing a new procedure for reward and penalty scheme in performance-based regulation of electricity distribution companies. *Int. Trans. Electr. Energ. Syst.* **2018**, *28*, e2628. [CrossRef]

12. Khonakdar-Tarsi, I.; Fotuhi-Firuzabad, M.; Ehsan, M.; Mohammadnezhad-Shourkaei, H.; Jooshaki, M. Reliability incentive regulation based on reward-penalty mechanism using distribution feeders clustering. *Int. Trans. Electr. Energ. Syst.* **2021**, e12958. [CrossRef]

13. Jooshaki, M.; Abbaspour, A.; Fotuhi-Firuzabad, M.; Moeini-Aghtaie, M.; Ozdemir, A. A new reward-penalty mechanism for distribution companies based on concept of competition. In Proceedings of the IEEE PES Innovative Smart Grid Technologies, Istanbul, Turkey, 12–15 October 2014.

14. Mozaffari, M.; Abyaneh, A.H.; Jooshaki, M.; Moeini-Aghtaie, M. Joint expansion planning studies of EV parking lots placement and distribution network. *IEEE Trans. Ind. Inform.* **2020**, *16*, 6455–6465.

15. Jooshaki, M.; Abbaspour, A.; Fotuhi-Firuzabad, M.; Moeini-Aghtaie, M.; Lehtonen, M. Incorporating the effects of service quality regulation in decision-making framework of distribution companies. *IET Gener. Transm. Distrib.* **2018**, *12*, 4172–4181. [CrossRef]

16. Jooshaki, M.; Farzin, H.; Abbaspour, A.; Fotuhi-Firuzabad, M.; Lehtonen, M. A risk-based framework to optimize distributed generation investment plans considering incentive reliability regulations. In Proceedings of the International Conference and Exhibition on Electricity Distribution (CIRED), Madrid, Spain, 3 June 2019.

17. Hyder, Z.; Siau, K.; Nah, F. Artificial intelligence, machine learning, and autonomous technologies in mining industry. *J. Database Manag.* **2019**, *30*, 67–79. [CrossRef]

18. Ali, D.; Frimpong, S. Artificial intelligence, machine learning and process automation: Existing knowledge frontier and way forward for mining sector. *Artif. Intell. Rev.* **2020**, *53*, 6025–6042. [CrossRef]

19. Michaux, S. *Global Outlook for Graphite, GTK Internal Report 2018a*; Geological Survey of Finland (GTK): Espoo, Finland, 2018.

20. Michaux, S. *Global Outlook for Magnesium Metal, GTK Internal Report 2018b*; Geological Survey of Finland (GTK): Espoo, Finland, 2018.

21. Michaux, S. Oil from a Critical Raw Material Perspective. In GTK Open File Work Report 2019. Serial Number 70/2019; ISBN 978-952-217-404-8. Available online: http://tupa.gtk.fi/raportti/arkisto/70_2019.pdf (accessed on 6 July 2021).

22. Michaux, S. *Global Outlook for Tungsten, GTK Open File Work Report*; Geological Survey of Finland (GTK): Espoo, Finland, 2021.

23. Petruk, W. *Applied Mineralogy in the Mining Industry*, 1st ed.; Elsevier: Amsterdam, The Netherlands, 2000; pp. 1–2.

24. Avalos, S.; Kracht, W.; Ortiz, J.M. Machine learning and deep learning methods in mining operations: A data-driven SAG mill energy consumption prediction application. *Min. Metall. Explor.* **2020**, *37*, 1197–1212. [CrossRef]

25. Inapakurthi, R.K.; Miriyala, S.S.; Mitra, K. Recurrent neural networks based modelling of industrial grinding operation. *Chem. Eng. Sci.* **2020**, *219*, 115585. [CrossRef]

26. Rodriguez-Galiano, V.; Sanchez-Castillo, M.; Chica-Olmo, M.; Chica-Rivas, M. Machine learning predictive models for mineral prospectivity: An evaluation of neural networks, random forest, regression trees and support vector machines. *Ore Geol. Rev.* **2015**, *71*, 804–818. [CrossRef]

27. Lewkowski, C.; Porwal, A.; González-Álvarez, I. Genetic programming applied to base-metal prospectivity mapping in the Aravalli Province, India. In Proceedings of the EGU General Assembly Conference Abstracts, Vienna, Austria, 2–7 May 2010; p. 523.

28. Leite, E.P.; Souza Filho, C.R. Artificial neural networks applied to mineral potential mapping for copper-gold mineralizations in the Carajás Mineral Province, Brazil. *Geophys. Prospect.* **2009**, *57*, 1049–1065. [CrossRef]

29. Zuo, R.; Xiong, Y. Big data analytics of identifying geochemical anomalies supported by machine learning methods. *Nat. Resour. Res.* **2018**, *27*, 5–13. [CrossRef]

30. Zuo, R. Machine learning of mineralization-related geochemical anomalies: A review of potential methods. *Nat. Resour. Res.* **2017**, *26*, 457–464. [CrossRef]

31. Wang, Z.; Zuo, R.; Dong, Y. Mapping geochemical anomalies through integrating random forest and metric learning methods. *Nat. Resour. Res.* **2019**, *28*, 1285–1298. [CrossRef]

32. Matthew, J.C.; Anya, M.R. Geological mapping using remote sensing data: A comparison of five machine learning algorithms, their response to variations in the spatial distribution of training data and the use of explicit spatial information. *Comput. Geosci.* **2014**, *63*, 22–33.

33. Harvey, A.S.; Fotopoulos, G. Geological mapping using machine learning algorithms. *Int. Arch. Photogramm. Remote. Sens. Spat. Inf. Sci.* **2016**, *41*, 423–430. [CrossRef]

34. Radford, D.D.; Cracknell, M.J.; Roach, M.J.; Cumming, G.V. Geological mapping in western Tasmania using radar and random forests. *IEEE J. Sel. Top. Appl. Earth Obs. Remote. Sens.* **2018**, *11*, 3075–3087. [CrossRef]

35. Lhissou, R.; El Harti, A.; Maimouni, S.; Adiri, Z. Assessment of the image-based atmospheric correction of multispectral satellite images for geological mapping in arid and semi-arid regions. *Remote Sens. Appl. Soc. Environ.* **2020**, *20*, 100420.

36. Barker, R.D.; Barker, S.L.; Cracknell, M.J.; Stock, E.D.; Holmes, G. Quantitative Mineral Mapping of Drill Core Surfaces II: Long-Wave Infrared Mineral Characterization Using μ XRF and Machine Learning. *Econ. Geol.* **2021**, *116*, 821–836. [CrossRef]

37. Acosta, I.C.C.; Khodadadzadeh, M.; Tolosana-Delgado, R.; Gloaguen, R. Drill-core hyperspectral and geochemical data integration in a Superpixel-based machine learning framework. *IEEE J. Sel. Top. Appl. Earth Obs. Remote Sens.* **2020**, *13*, 4214–4228. [CrossRef]

38. Contreras, C.; Khodadadzadeh, M.; Ghamisi, P.; Gloaguen, R. Mineral mapping of drill core hyperspectral data with extreme learning machines. In Proceedings of the 2019 IEEE International Geoscience and Remote Sensing Symposium (IGARSS), Yokohama, Japan, 28 July–2 August 2019.

39. Chauhan, S.; Rühaak, W.; Khan, F.; Enzmann, F.; Mielke, P.; Kersten, M.; Sass, I. Processing of rock core microtomography images: Using seven different machine learning algorithms. *Comput. Geosci.* **2016**, *86*, 120–128. [CrossRef]

40. Wang, Y.; Lin, C.L.; Miller, J.D. Improved 3D image segmentation for X-ray tomographic analysis of packed particle beds. *Miner. Eng.* **2015**, *83*, 185–191. [CrossRef]

41. Guntoro, P.I.; Ghorbani, Y.; Butcher, A.R.; Kuva, J.; Rosenkranz, J. Textural Quantification and Classification of Drill Cores for Geometallurgy: Moving Toward 3D with X-ray Microcomputed Tomography (μCT). *Nat. Resour. Res.* **2020**, *29*, 3547–3565. [CrossRef]

42. Page, M.J.; Moher, D.; Bossuyt, P.M.; Boutron, I.; Hoffmann, T.C.; Mulrow, C.D.; Shamseer, L.; Tetzlaff, J.M.; AKI, E.A.; Brennan, S.E.; et al. PRISMA 2020 explanation and elaboration: Updated guidance and exemplars for reporting systematic reviews. *BMJ* **2021**, *372*, 1–36.

43. Jung, D.; Choi, Y. Systematic review of machine learning applications in mining: Exploration, exploitation, and reclamation. *Minerals* **2021**, *11*, 148. [CrossRef]

44. Scopus. Available online: https://www.scopus.com/home.uri (accessed on 1 July 2021).

45. Luo, Z.; Xiong, Y.; Zuo, R. Recognition of geochemical anomalies using a deep variational autoencoder network. *Appl. Geochem.* **2020**, *122*, 104710. [CrossRef]

46. Xiong, Y.; Zuo, R. Recognition of geochemical anomalies using a deep autoencoder network. *Comput. Geosci.* **2016**, *86*, 75–82. [CrossRef]

47. Wang, C.; Pan, Y.; Chen, J.; Ouyang, Y.; Rao, J.; Jiang, Q. Indicator element selection and geochemical anomaly mapping using recursive feature elimination and random forest methods in the Jingdezhen region of Jiangxi Province, South China. *Appl. Geochem.* **2020**, *122*, 104760. [CrossRef]

48. Lawley, C.J.; Somers, A.M.; Kjarsgaard, B.A. Rapid geochemical imaging of rocks and minerals with handheld laser induced breakdown spectroscopy (LIBS). *J. Geochem. Explor.* **2021**, *222*, 106694. [CrossRef]

49. Maepa, F.; Smith, R.S.; Tessema, A. Support Vector Machine and Artificial Neural Network Modelling of Orogenic Gold Prospectivity Mapping in the Swayze greenstone belt, Ontario, Canada. *Ore Geol. Rev.* **2020**, *130*, 103968. [CrossRef]

50. McMillan, M.; Haber, E.; Peters, B.; Fohring, J. Mineral prospectivity mapping using a VNet convolutional neural network. *Lead. Edge* **2021**, *40*, 99–105. [CrossRef]

51. Li, H.; Li, X.; Yuan, F.; Jowitt, S.M.; Zhang, M.; Zhou, J.; Wu, B. Convolutional neural network and transfer learning based mineral prospectivity modeling for geochemical exploration of Au mineralization within the Guandian–Zhangbaling area, Anhui Province, China. *Appl. Geochem.* **2020**, *122*, 104747. [CrossRef]

52. Granek, J.; Haber, E. Data mining for real mining: A robust algorithm for prospectivity mapping with uncertainties. In Proceedings of the 2015 SIAM International Conference on Data Mining. Society for Industrial and Applied Mathematics, Vancouver, BC, Canada, April 30–May 2 2015; pp. 145–153.

53. Schneider, S.; Murphy, R.J.; Melkumyan, A.; Nettleton, E. Autonomous mapping of mine face geology using hyperspectral data. In Proceedings of the 35th APCOM Symp, Wollongong, Australia, 24–30 September 2011; pp. 24–30.

54. Xiong, Y.; Zuo, R. Recognizing multivariate geochemical anomalies for mineral exploration by combining deep learning and one-class support vector machine. *Comput. Geosci.* **2020**, *140*, 104484. [CrossRef]

55. Wang, Z.; Dong, Y.; Zuo, R. Mapping geochemical anomalies related to Fe–polymetallic mineralization using the maximum margin metric learning method. *Ore Geol. Rev.* **2019**, *107*, 258–265. [CrossRef]

56. Wang, J.; Zhou, Y.; Xiao, F. Identification of multi-element geochemical anomalies using unsupervised machine learning algorithms: A case study from Ag–Pb–Zn deposits in north-western Zhejiang, China. *Appl. Geochem.* **2020**, *120*, 104679. [CrossRef]

57. Chakouri, M.; El Harti, A.; Lhissou, R.; El Hachimi, J.; Jellouli, A. Geological and Mineralogical mapping in Moroccan central Jebilet using multispectral and hyperspectral satellite data and Machine Learning. *Int. J.* **2020**, *9*, 5772–5783. [CrossRef]

58. Roshanravan, B. Translating a mineral systems model into continuous and data-driven targeting models: An example from the Dolatabad chromite district, southeastern Iran. *J. Geochem. Explor.* **2020**, *215*, 106556. [CrossRef]

59. Bérubé, C.L.; Olivo, G.R.; Chouteau, M.; Perrouty, S.; Shamsipour, P.; Enkin, R.J.; Thiémonge, R. Predicting rock type and detecting hydrothermal alteration using machine learning and petrophysical properties of the Canadian Malartic ore and host rocks, Pontiac Subprovince, Québec, Canada. *Ore Geol. Rev.* **2018**, *96*, 130–145. [CrossRef]

60. Schneider, S.; Melkumyan, A.; Murphy, R.J.; Nettleton, E. Classification of hyperspectral imagery using GPs and the OAD covariance function with automated endmember extraction. In Proceedings of the 2011 IEEE 23rd International Conference on Tools with Artificial Intelligence, Boca Raton, FL, USA, 7–9 November 2011; pp. 579–584.

61. Gloaguen, R.; Kirsch, M.; Lorenz, S.; Booysen, R.; Zimmermann, R.; Ghamisi, P.; Rasti, B. Towards 4D Virtual Outcrops with Hyperspectral Imaging. In Proceedings of the IGARSS 2020-2020 IEEE International Geoscience and Remote Sensing Symposium, Waikoloa, HI, USA, 26 September–2 October 2020; pp. 4035–4036.

62. Jackisch, R.; Lorenz, S.; Kirsch, M.; Zimmermann, R.; Tusa, L.; Pirttijärvi, M.; Gloaguen, R. Integrated Geological and Geophysical Mapping of a Carbonatite-Hosting Outcrop in Siilinjärvi, Finland, Using Unmanned Aerial Systems. *Remote Sens.* **2020**, *12*, 2998. [CrossRef]

63. Tuşa, L.; Khodadadzadeh, M.; Contreras, C.; Rafiezadeh Shahi, K.; Fuchs, M.; Gloaguen, R.; Gutzmer, J. Drill-Core Mineral Abundance Estimation Using Hyperspectral and High-Resolution Mineralogical Data. *Remote Sens.* **2020**, *12*, 1218. [CrossRef]

64. Contreras, I.C.; Khodadadzadeh, M.; Gloaguen, R. Multi-Label Classification for Drill-Core Hyperspectral Mineral Mapping. *Int. Arch. Photogramm. Remote. Sens. Spat. Inf. Sci.* **2020**, *43*, 383–388. [CrossRef]

65. Lorenz, S.; Seidel, P.; Ghamisi, P.; Zimmermann, R.; Tusa, L.; Khodadadzadeh, M.; Gloaguen, R. Multi-sensor spectral imaging of geological samples: A data fusion approach using spatio-spectral feature extraction. *Sensors* **2019**, *19*, 2787. [CrossRef]

66. Koch, P.H.; Lund, C.; Rosenkranz, J. Automated drill core mineralogical characterization method for texture classification and modal mineralogy estimation for geometallurgy. *Miner. Eng.* **2019**, *136*, 99–109. [CrossRef]

67. Contreras, C.; Khodadadzadeh, M.; Tusa, L.; Ghamisi, P.; Gloaguen, R. A machine learning technique for drill core hyperspectral data analysis. In Proceedings of the 9th Workshop on Hyperspectral Image and Signal Processing: Evolution in Remote Sensing (WHISPERS), Amsterdam, The Netherlands, 23–26 September 2018; pp. 1–5.

68. Acosta, I.C.C.; Khodadadzadeh, M.; Tusa, L.; Ghamisi, P.; Gloaguen, R. A machine learning framework for drill-core mineral mapping using hyperspectral and high-resolution mineralogical data fusion. *IEEE J. Sel. Top. Appl. Earth Obs. Remote Sens.* **2019**, *12*, 4829–4842. [CrossRef]

69. Wills, B.A.; Napier-Munn, T.J. *Mineral Processing Technology: An Introduction to the Practical Aspects of Ore Treatment and Mineral Recovery*, 7th ed.; Butterworth-Heinemann: Burlington, MA, USA, 2006; pp. 7–8.

70. Leroy, S.; Pirard, E. Mineral recognition of single particles in ore slurry samples by means of multispectral image processing. *Miner. Eng.* **2019**, *132*, 228–237. [CrossRef]

71. Coelho, P.A.; Sandoval, C.; Alvarez, J.; Sanhueza, I.; Godoy, C.; Torres, S.; Sbarbaro, D. Automatic near-infrared hyperspectral image analysis of copper concentrates. *IFAC-PapersOnLine* **2019**, *52*, 94–98. [CrossRef]

72. Karimpouli, S.; Tahmasebi, P. Segmentation of digital rock images using deep convolutional autoencoder networks. *Comput. Geosci.* **2019**, *126*, 142–150. [CrossRef]

73. Guntoro, P.I.; Tiu, G.; Ghorbani, Y.; Lund, C.; Rosenkranz, J. Application of machine learning techniques in mineral phase segmentation for X-ray microcomputed tomography (μCT) data. *Miner. Eng.* **2019**, *142*, 105882. [CrossRef]

74. Shipman, W.J.; Nel, A.L.; Chetty, D.; Miller, J.D.; Lin, C.L. The application of machine learning to the problem of classifying voxels in X-ray microtomographic scans of mineralogical samples. In Proceedings of the 2013 IEEE International Conference on Industrial Technology (ICIT), Cape Town, South Africa, 25–28 February 2013; pp. 1184–1189.

75. Domínguez-Olmedo, J.L.; Toscano, M.; Mata, J. Application of classification trees for improving optical identification of common opaque minerals. *Comput. Geosci.* **2020**, *140*, 104480. [CrossRef]

76. Li, X.; Zhang, C.; Behrens, H.; Holtz, F. Calculating amphibole formula from electron microprobe analysis data using a machine learning method based on principal components regression. *Lithos* **2020**, *362*, 105469. [CrossRef]

77. Borges, H.P.; de Aguiar, M.S. Mineral classification using machine learning and images of microscopic rock thin section. In Proceedings of the Mexican International Conference on Artificial Intelligence, Xalapa, Mexico, 28 October–1 November 2019; Springer: Berlin/Heidelberg, Germany, 2019.

78. Gomes, O.D.F.M.; Iglesias, J.C.A.; Paciornik, S.; Vieira, M.B. Classification of hematite types in iron ores through circularly polarized light microscopy and image analysis. *Miner. Eng.* **2013**, *52*, 191–197. [CrossRef]

79. Silva, C.A.M.; Ellefmo, S.L.; Sandøy, R.; Sørensen, B.E.; Aasly, K. A neural network approach for spatial variation assessment–A nepheline syenite case study. *Miner. Eng.* **2020**, *149*, 106178. [CrossRef]

80. Iglesias, J.C.Á.; Santos, R.B.M.; Paciornik, S. Deep learning discrimination of quartz and resin in optical microscopy images of minerals. *Miner. Eng.* **2019**, *138*, 79–85. [CrossRef]

81. Maepa, F.M.; Smith, R.S. Radial basis function link nets method for predicting gold mineral potential from geological and geophysical data in the Swayze greenstone belt (SGB). In *SEG Technical Program Expanded Abstracts 2018, Society of Exploration Geophysicists*; Society of Exploration Geophysicists: Tulsa, OK, USA, 2018; pp. 1848–1852.

82. Chen, S.; Hattori, K.; Grunsky, E.C. Identification of sandstones above blind uranium deposits using multivariate statistical assessment of compositional data, Athabasca Basin, Canada. *J. Geochem. Explor.* **2018**, *188*, 229–239. [CrossRef]

83. Chen, S.; Plouffe, A.; Hattori, K. A multivariate statistical approach identifying the areas underlain by potential porphyry-style Cu mineralization, south-central British Columbia, Canada. *J. Geochem. Explor.* **2019**, *202*, 13–26. [CrossRef]

84. Chan, S.A.; Hassan, A.M.; Humphrey, J.D.; Mahmoud, M.A.; Abdulraheem, A. Evaluation of brittleness index based on mineral compositions prediction using artificial neural network. In Proceedings of the ARMA/DGS/SEG International Geomechanics Symposium, American Rock Mechanics Association. 3–5 November 2020; Virtual event. Available online: https://seg.org/Events/International-Geomechanics-Symposium (accessed on 19 July 2021).

85. Krutko, V.; Belozerov, B.; Budennyy, S.; Sadikhov, E.; Kuzmina, O.; Orlov, D.; Koroteev, D. A New approach to clastic rocks pore-scale topology reconstruction based on automatic thin-section images and CT scans analysis. In Proceedings of the SPE Annual Technical Conference and Exhibitionm, Alberta, AB, Canada, 23 September 2019.

86. Schneider, S.; Murphy, R.J.; Melkumyan, A. Evaluating the performance of a new classifier–the GP-OAD: A comparison with existing methods for classifying rock type and mineralogy from hyperspectral imagery. *ISPRS J. Photogramm. Remote Sens.* **2014**, *98*, 145–156. [CrossRef]

87. Ksieniewicz, P.; Graña, M.; Woźniak, M. Blurred labeling segmentation algorithm for hyperspectral images. *Comput. Collect. Intell. Lect. Notes Comput. Sci.* **2015**, *9330*, 578–587.

88. Melkumyan, A.; Murphy, R.J. Spectral domain noise suppression in dual-sensor hyperspectral imagery using Gaussian processes. In Proceedings of the International Conference on Neural Information Processing, Vancouver, BC, Canada, 22–25 November 2010; Springer: Berlin/Heidelberg, Germany, 2010.

89. Diaz, G.F.; Ortiz, J.M.; Silva, J.F.; Lobos, R.A.; Egana, A.F. Variogram-Based Descriptors for Comparison and Classification of Rock Texture Images. *Math. Geosci.* **2020**, *52*, 451–476. [CrossRef]

90. Camalan, M.; Çavur, M.; Hoşten, Ç. Assessment of chromite liberation spectrum on microscopic images by means of a supervised image classification. *Powder Technol.* **2017**, *322*, 214–225. [CrossRef]

91. Chen, L.; Guan, Q.; Feng, B.; Yue, H.; Wang, J.; Zhang, F. A multi-convolutional autoencoder approach to multivariate geochemical anomaly recognition. *Minerals* **2019**, *9*, 270. [CrossRef]

92. Ordóñez-Calderón, J.C.; Gelcich, S. Machine learning strategies for classification and prediction of alteration facies: Examples from the Rosemont Cu-Mo-Ag skarn deposit, SE Tucson Arizona. *J. Geochem. Explor.* **2018**, *194*, 167–188. [CrossRef]

93. Hood, S.B.; Cracknell, M.J.; Gazley, M.F. Linking protolith rocks to altered equivalents by combining unsupervised and supervised machine learning. *J. Geochem. Explor.* **2018**, *186*, 270–280. [CrossRef]

94. Bajwa, I.S.; Choudhary, M.A. A study for prediction of minerals in rock images using back propagation neural networks. In Proceedings of 2006 International Conference on Advances in Space Technologies, Islamabad, Pakistan, 2–3 September 2006.

95. Gallagher, M.; Deacon, P. Neural networks and the classification of mineralogical samples using X-ray spectra. In Proceedings of the 9th International Conference on Neural Information Processing, Singapore, 18–22 November 2002.

96. Ishikawa, S.T.; Gulick, V.C. An automated mineral classifier using Raman spectra. *Comput. Geosci.* **2013**, *54*, 259–268. [CrossRef]

97. Carr, D.A.; Lach-hab, M.; Yang, S.; Vaisman, I.I.; Blaisten-Barojas, E. Machine learning approach for structure-based zeolite classification. *Microporous Mesoporous Mater.* **2009**, *117*, 339–349. [CrossRef]

98. Van Duijvenbode, J.R.; Buxton, M.W.; Shishvan, M.S. Performance improvements during mineral processing using material fingerprints derived from machine learning—A conceptual framework. *Minerals* **2020**, *10*, 366. [CrossRef]

99. Metelka, V.; Baratoux, L.; Jessell, M.W.; Barth, A.; Ježek, J.; Naba, S. Automated regolith landform mapping using airborne geophysics and remote sensing data, Burkina Faso, West Africa. *Remote Sens. Environ.* **2018**, *204*, 964–978. [CrossRef]

100. Ramsey, J.; Gazis, P.; Roush, T.; Spirtes, P.; Glymour, C. Automated remote sensing with near infrared reflectance spectra: Carbonate recognition. *Data Min. Knowl. Discov.* **2002**, *6*, 277–293. [CrossRef]

101. Rahman, A.; Timms, G.; Shahriar, M.S.; Sennersten, C.; Davie, A.; Lindley, C.A.; Coombe, M. Association between imaging and XRF sensing: A machine learning approach to discover mineralogy in abandoned mine voids. *IEEE Sens. J.* **2016**, *16*, 4555–4565. [CrossRef]

102. Pradhan, A. Support vector machine-a survey. *Int. J. Emerg. Technol. Adv. Eng.* **2012**, *2*, 82–85.

103. Boulesteix, A.L.; Janitza, S.; Kruppa, J.; König, I.R. Overview of random forest methodology and practical guidance with emphasis on computational biology and bioinformatics. *Wiley Interdiscip. Rev. Data Min. Knowl. Discov.* **2012**, *2*, 493–507. [CrossRef]

104. Goodfellow, I.; Bengio, Y.; Courville, A. Deep Learning. MIT Press: Massachusetts, MA, USA, 2016; pp. 1–26.

105. Kanungo, T.; Mount, D.M.; Netanyahu, N.S.; Piatko, C.D.; Silverman, R.; Wu, A.Y. An efficient k-means clustering algorithm: Analysis and implementation. *IEEE Trans. Pattern Anal. Mach. Intell.* **2002**, *24*, 881–892. [CrossRef]

106. Murtagh, F.; Contreras, P. Algorithms for hierarchical clustering: An overview. *Wiley Interdiscip. Rev. Data Min. Knowl. Discov.* **2012**, *2*, 86–97. [CrossRef]

107. Barroso, A.R.F.; Baiden, G.; Johnson, J. Knowledge representation and expert systems for mineral processing using Infobright. In Proceedings of the 2010 IEEE International Conference on Granular Computing, Brussels, Belgium, 14–16 August 2010.

MDPI
St. Alban-Anlage 66
4052 Basel
Switzerland
Tel. +41 61 683 77 34
Fax +41 61 302 89 18
www.mdpi.com

Minerals Editorial Office
E-mail: minerals@mdpi.com
www.mdpi.com/journal/minerals